A Collection of Questions and Problems in Physics

D1455177

Л. А. Сена

Сборник вопросов и задач по физике

Издательство «Высшая школа»
Москва

L. A. Sena

A Collection
of Questions and Problems
in Physics

Mir Publishers Moscow

CBS Publishers & Distributors Pvt. Ltd.
New Delhi • Bengaluru • Chennai • Kochi • Kolkata • Mumbai
Hyderabad • Nagpur • Patna • Pune • Vijayawada

Translated from the Russian
by Eugene Yankovsky

First published 1988
Revised from the 1986 Russian edition

УДК 53(075.8)=20

Учебное пособие

Лев Аронович Сена

СБОРНИК ВОПРОСОВ И ЗАДАЧ ПО ФИЗИКЕ

© Издательство «Высшая школа», 1986
© English translation, Mir Publishers,
  1988

CBS Pubs ISBN:81-239-0305-7
Mir Pubs ISBN:5-03-000125-5

First Indian Edition: 1994
Reprint: 2001, 2002, 2004

Copyright © English Translation, Mir Publishers, 1988

This edition has been published in India by arrangement with
Mir Publishers, Moscow

All rights reserved. No part of this book may be reproduced or transmitted in any
form or by any means, electronic or mechanical, including photocopying,
recording, or any information storage and retrieval system without permission, in
writing, from the publisher.

Published by **Satish Kumar Jain** and produced by **Varun Jain** for

CBS Publishers & Distributors Pvt. Ltd.,
4819/XI Prahlad Street, 24 Ansari Road, Daryaganj, New Delhi - 110002
delhi@cbspd.com, cbspubs@airtelmail.in • www.cbspd.com

Corporate Office: 204 FIE, Industrial Area, Patparganj, Delhi - 110 092
Ph: 49344934 • Fax: 011-49344935
E-mail: publishing@cbspd.com • publicity@cbspd.com

Branches:
• *Bengaluru:* 2975, 17th Cross, K.R. Road, Bansankari 2nd Stage,
  Bengaluru - 70 • Ph: +91-80-26771678/79 • Fax: +91-80-26771680
• *Chennai:* No. 7, Subbaraya Street, Shenoy Nagar, Chennai - 600030
  Ph: +91-44-26681266, 26680620 • Fax: +91-44-42032115
• *Kochi:* Ashana House, 39/1904, A.M. Thomas Road, Valanjambalam,
  Ernakulum, Kochi • Ph: +91-484-4059061-65
• *Kolkata:* 6-B, Ground Floor, Rameshwar Shaw Road, Kolkata - 700014
  Ph: +91-33-22891126/7/8 • E-mail: kolkata@cbspd.com
• *Mumbai:* 83-C, Dr. E. Moses Road, Worli, Mumbai - 400018
  Ph: +91-9833017933, 022-24902340/41 • E-mail: mumbai@cbspd.com

Printed at: J.S. Offset Printers, Delhi (India)

Preface

To solve the problems that modern science and technology pose, specialists must not only possess a certain volume of knowledge but must be able to freely apply this knowledge. The aim of the present collection of questions and problems is to develop practical skills during study of one of the fundamental sciences, physics. The *Collection* is intended for the self-instruction of students of technical colleges. The best way to use it is to solve the problems while preparing for term exams.

The *Collection* contains more than 400 questions and problems covering all the sections of the physics course. All questions and problems have detailed answers and solutions. For this reason the two main sections of the book, Questions and Problems and Answers and Solutions, have identical headings and numbering: each chapter in the first section has a corresponding chapter in the second, and the numbering of answers corresponds to the numbering of problems.

A special feature of the *Collection* is the drawings and diagrams for most of the questions and answers. The diagrams use a variety of scales: linear, semilog, log-log, and quadratic.

Arrangement of the material in this *Collection* corresponds to the structure most commonly used in college physics textbooks. One exception is the questions and problems involving the special theory of relativity. These are placed in different chapters, starting from the one dealing with mechanics.

While preparing the manuscript, I received many suggestions and comments from institutions of higher learning in Leningrad, Moscow, and Tomsk. I take this opportunity to thank all who helped to improve this book. I am particularly grateful to Professors I. A. Yakovlev, B. M. Smirnov, V. A. Fabrikant, and S. Ya. Shats. I would also like to thank Prof. A. G. Chertov and the Department of General Physics at the Moscow Physical Engineering Institute for most useful comments offered while reviewing this book.

L. A. Sena

5

Contents

> A drawing is the source and soul
> of every image and the root of every
> science."
>
> *Michelangelo*

Introduction

The student put down his record book and picked up an examination card. Upon reading it, he gasped: "My God! What will I do?" Judging by his face, one would think he held at the very least a poisonous snake. The assignment on the card read: "The velocity distribution of molecules; the Maxwellian curve." The student was not required to derive the formula or even write out the formula. All he had to do was to draw the curve and explain its physical meaning. Another student, in drawing the van der Waals isotherm depicted something resembling a camel with two humps; moreover, the curve passed through the origin. Still another student, while explaining the idea behind the Stern experiment, made the outer cylinder rotate while the inner cylinder remained fixed. Finally, to the question of how the temperature of a gas changes under adiabatic expansion a student gave the following "reasonable" answer: since objects expand when heated, and the gas expanded in the experiment, the temperature must have risen.

Unfortunately, examples of such answers abound. We are not speaking of the excellent student or even of the average student, of course. Yet it can be said without exaggeration that for many students "qualitative" questions and problems present many more difficulties than the derivation of formulas. The situation is especially bad with the building of diagrams and sketching of experiments. College instructors and lecturers often complain of the low level of school instruction, but complaints are of no help. Hence, it is essential to develop a student's creative thinking and ability to analyze physical phenomena.

It was this that prompted me to draw on more than a half-century of instruction at colleges in Leningrad and compile the present collection of questions and problems.

The book was conceived literally as a teaching aid; it is intended to help the student in the physics course at the freshman level. The main emphasis is on the use of diagrams and sketches. A drawing makes the essence of a problem clearer and assists the development of "qualitative" thinking. That is why I have chosen Michelangelo's remarkable words for the epigraph to this work.

The questions and problems found here encompass practically all sections of the physics course studied in a technical college. Since some colleges give greater stress to certain topics, the book includes a number of questions and problems intended for a well-prepared student. This feature makes it possible to use the book to some extent in the physics departments of universities and the physics and mathematics departements of teachers' colleges. On the other hand, some problems require only knowledge within the scope of secondary school, though these are usually not considered in the school syllabus in such form.

All questions and problems have detailed answers and solutions. At times a variety of solutions are given. One may be based on dimensionality considerations, while another is achieved through direct integration. The majority of answers and solutions are analyzed and discussed. Sometimes practical applications are given to show how and where the specific phenomena and laws are encountered.

In compiling this collection I did not aim at selecting the most difficult or the least difficult questions and problems. The range of difficulty is considerable. The book is structured in such a way that all students, from the well-prepared to the not-so-well-prepared, can use it. If a student is not able to answer a question or solve a problem without help, a careful study of the solution will help him to master the theory involved and solve on his own at first the simpler problems and then the more complicated. A well-prepared student will be able to solve most of the problems, but even he will find it helpful to compare his solution with the one given in the book and to read the accompanying discussion of the results.

To answer the questions it is sufficient at times to read the question, glance at the diagram, and write the appropriate formula. On the other hand, some problems require constructing a diagram or even reconstructing the diagram accompanying the problem. Others necessitate making simple mathematical transformations, still others solving the problem in general form, using the necessary concepts of mathematics.

In this connection the question of the role and necessary level of mathematical knowledge arises. I have assumed that what the student learns in the accompanying mathematical course may and must be employed when necessary. I object to what is jokingly called the "formulization" of physics, but I also object to ignoring the possibilities offered by mathematics. A knowledge of mathematics is essential for a study of special disciplines. And, vice versa, a study of these disciplines is extremely useful for a deeper understanding of mathematical concepts and methods. Bearing all this in mind, I have set as the "upper limit" the use of the most simple ordinary differential equations of an order no higher than the second.

Notwithstanding the great convenience of the symbolic method in the theory of oscillations and the theory of alternating currents, the respective problems have been solved by the common trigonometric method with occasional employment of the vector concept. This is done for the simple reason that the symbolic method is not studied in the course of general physics in most technical colleges, and justifiably, I believe, because for first-year students the method is too formal and lacks pictorial clarity.

Questions and Problems

1. Fundamentals of Mechanics

1.1. A wind is blowing with a constant velocity v in the direction denoted by the arrow in the figure. Two airplanes start out from a point A and fly with a constant speed c. One flies against the wind to a point B and then returns to point A, while the other flies in the direction perpendicular to the wind to a point C and then returns to point A. The distances AB and AC are the same.

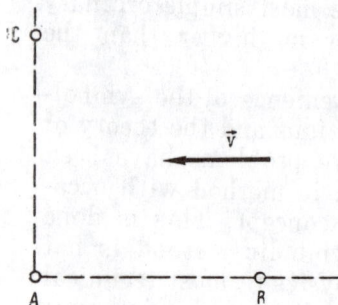

Fig. 1.1

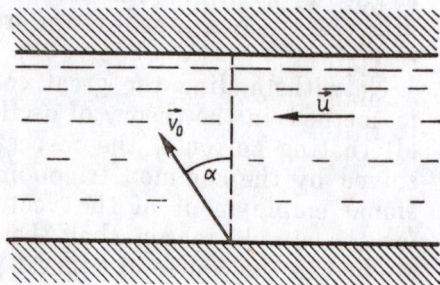

Fig. 1.2

Which plane will return to point A first and what will be the ratio of the flight times of the two planes?

1.2. A boat is moving across a river whose waters flow with a velocity u. The velocity of the boat with respect to the current, v_0, is directed at an angle α to the line perpendicular to the current. What will be the angle θ at which the boat moves with respect to this line? What will be the velocity v of the boat with respect to the river banks? What should be the angle at which the boat moves directly across the current with given u and v?

1.3. From a point A on a bank of a channel with still waters a person must get to a point B on the opposite bank. All the distances are shown in the figure. The person uses a boat to travel across the channel and then walks along the bank to point B. The velocity of the boat is v_1 and the velocity of the walking person is v_2. Prove that the fastest way for the person to get from A

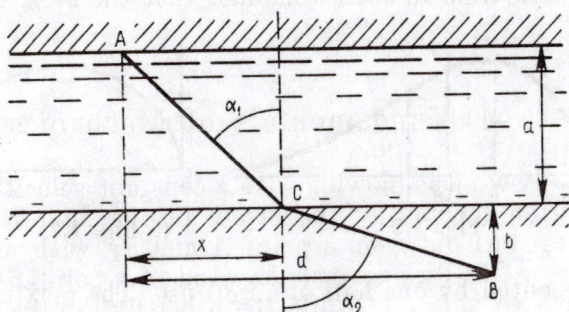

Fig. 1.3

to B is to select the angles α_1 and α_2 in such a manner that $(\sin \alpha_1 / \sin \alpha_2) = v_1/v_2$.

1.4. An object slides without friction down an inclined plane from a point B to a point C that is distant a from

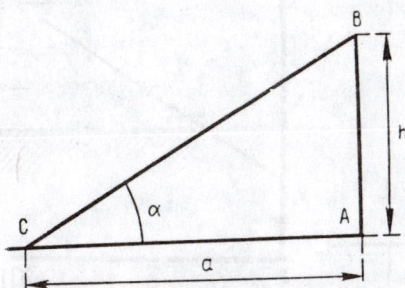

Fig. 1.4

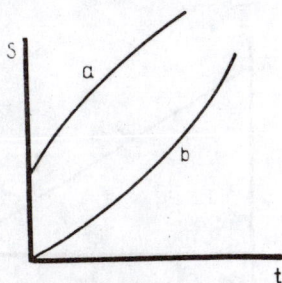

Fig. 1.5

a point A. At what height h (or at what angle α) is the sliding time minimal?

1.5. The time dependence of the lengths of the paths of two bodies moving in a straight line is given by curves a and b, respectively. What curve corresponds to accelerated motion and what curve to decelerated motion?

1.6. A material particle is moving along a straight line in such a manner that its velocity varies as shown in the figure. At which moment in time numbered successively on the time axis will the acceleration of the particle be maximal? How should one use the graph to determine the average velocity of motion over the time interval from t_1 to t_2?

1.7. The velocity of a particle moving in a straight line varies with time in such a manner that the v vs. t curve

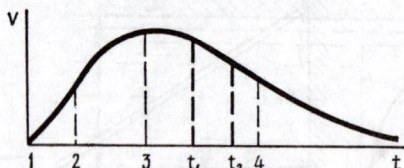

Fig. 1.6

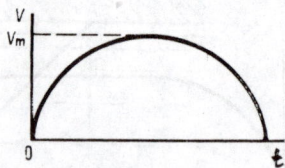

Fig. 1.7

is represented by one half of an ellipse. The maximal velocity is v_m and the total time of motion is t. What is the path traversed by the particle and the average velocity over t? Can such motion actually occur?

1.8. The velocity of a particle decreases in relation to the path traversed according to the linear law $v = v_0 - ax$. After what time will the particle get to a point B

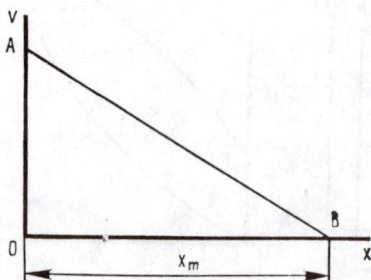

Fig. 1.8

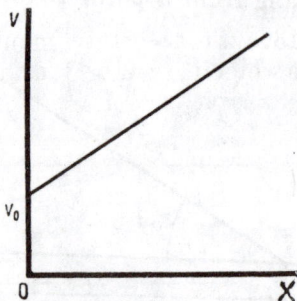

Fig. 1.9

that lies on the axis of abscissas distant x_m from the origin of coordinates?

1.9. The velocity of a particle moving in a straight line increases according to the linear law $v = v_0 + kx$. How does the acceleration change in the course of such motion? Does it increase or decrease or stay constant?

1.10. The figure shows the "timetable" of a train, the dependence of the speed of the train on the distance traveled. How can this graph be used to determine the average speed over the time interval it took the train to travel the entire distance?

1.11. A rod of length l leans by its upper end against a smooth vertical wall, while its other end leans against the floor. The end that leans against the wall moves uni-

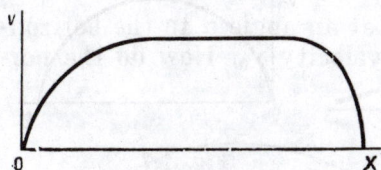

Fig. 1.10

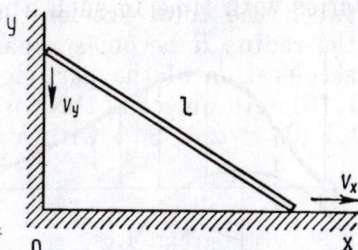

Fig. 1.11

formly downward. Will the other end move uniformly, too?

1.12. An object is thrown upward with an initial velocity v_0. The drag on the object is assumed to be proportional to the velocity. What time will it take the object to move upward and what maximal altitude will it reach?

1.13. At a certain moment in time the angle between the velocity vector $\mathbf{v}$ of a material particle and the acce-

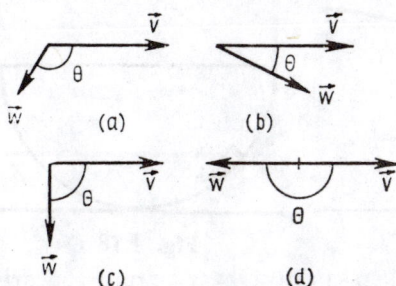

Fig. 1.13

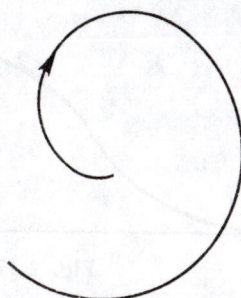

Fig. 1.14

leration vector $\mathbf{w}$ of that particle is θ. What will be the motion of the particle at this moment for different θ's: rectilinear or curvilinear, accelerated or uniform or decelerated?

1.14. A particle is moving along an expanding spiral in such a manner that the particle's normal acceleration remains constant. How will the linear and angular velocities change in the process?

1.15. A particle is moving in a circular orbit with a constant tangential acceleration. After a certain time t has elapsed after the beginning of motion, the angle between the total acceleration **w** and the direction along the radius R becomes equal to 45°. What is the angular acceleration of the particle?

1.16. An object is thrown at an angle α to the horizontal ($0° < \alpha < 90°$) with a velocity $\mathbf{v_0}$. How do the nor-

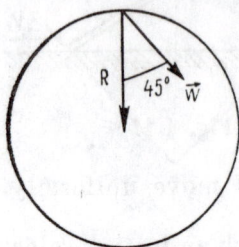

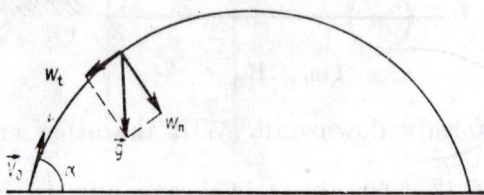

Fig. 1.15   **Fig. 1.16**

mal acceleration w_n and the tangential acceleration w_t vary in the process of ascent if the drag is ignored?

1.17. At the foot of a hill a certain velocity is imparted to a sled, as a result of which the sled moves up the hill

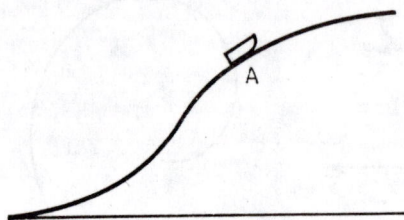

Fig. 1.17   **Fig. 1.18**

to a point A and then down the hill. What are the directions of the normal and tangential components of the acceleration at point A?

1.18. An object moves without friction along a concave surface. What are the directions of the normal and tangential components of the acceleration at the lowest possible point?

1.19. A stunt rider on a unicycle is riding around the arena of a circus in a circle of radius R. The radius of the wheel of the unicycle is r and the angular velocity with which the wheel rotates is ω. What is the angular acceleration of the wheel? (Ignore the fact that the wheel axis is inclined.)

1.20. A liquid has been poured into a cylindrical vessel of mass M (the mass of the vessel bottom can be ighored) and height H. The linear density of the liquid, that is, the ratio of the mass of the liquid column to its height, is δ.

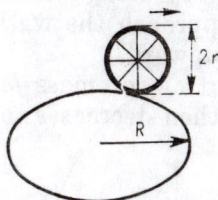

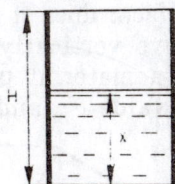

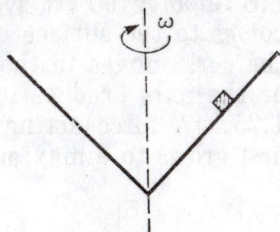

Fig. 1.19   Fig. 1.20   Fig. 1.21

What is the height x of the column of liquid at which the common center of gravity of the liquid plus the vessel is in the lowest position?

1.21. A cone-shaped funnel is being rotated with constant angular velocity ω. An object is placed on the inner

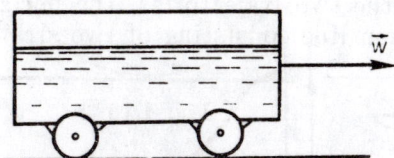

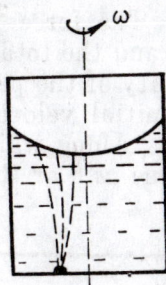

Fig. 1.22   Fig. 1.24

wall of the funnel. The object can freely move along the generatrix of the cone, but during the motion of the funnel the body is in a state of equilibrium. Is this equilibrium stable or unstable?

1.22. A vessel filled with water is moving horizontally with constant acceleration w. What shape will the surface of the liquid have?

15

1.23. A liquid has been poured into a cylindrical vessel. What shape will the surface of the liquid have if the vessel is rotated uniformly about its axis with an angular velocity ω?

1.24. A piece of cork has been attached to the bottom of a cylindical vessel that has been filled with water and is rotating about the vertical axis with a constant angular velocity ω. At some moment the cork gets free and comes to the surface. What is the trajectory along which the cork moves to the surface: does it approach the wall or the axis or does it move vertically upward?

1.25. A force acting on a material particle of mass m first grows to a maximum value F_m and then decreases to

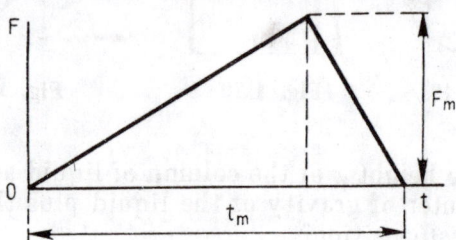

Fig. 1.25

zero. The force varies with time according to a linear law, and the total time of motion is t_m. What will be the velocity of the particle by the end of this time interval if the initial velocity is zero?

1.26. Along which of the two trajectories, the horizontal line $ac'b$ or the broken line consisting of two straight

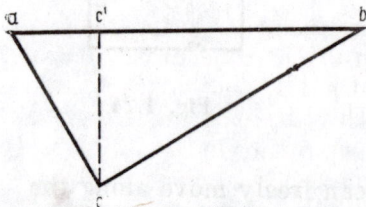

Fig. 1.26

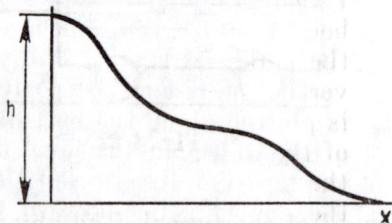

Fig. 1.27

segments (ac and cb), will the work performed by a force in displacing an object be greater if the friction is the same for all three straight segments?

16

1.27. An object of mass m is sliding down a hill of arbitrary shape and, after traveling a certain horizontal path, stops because of friction. The friction coefficient may be different for different segments of the entire path but it is independent of the velocity and direction of motion. Find the work that a force must perform to return the object to its initial position along the same path.

1.28. The dependence of the potential energy of an object on its position is given by the equation $W = ax^2$

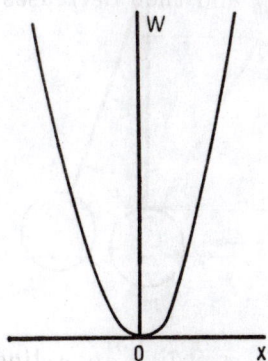

Fig. 1.28

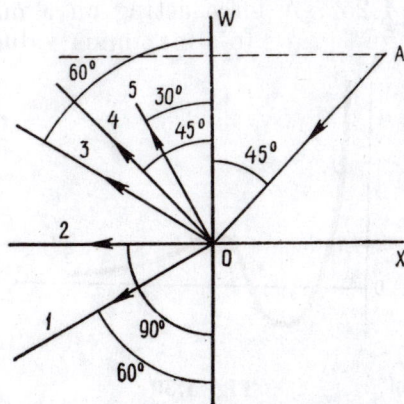

Fig. 1.29

(a parabola). What is the law by which the force acting on the object varies?

1.29. An object whose density is ρ_{ob} falls from a certain height into a liquid whose density is ρ_{liq}. In the figure the potential energy W of the object is plotted along the vertical axis and the position of the object (its altitude) is plotted along the horizontal axis. The potential energy of the object at the level of the liquid is taken zero and the positive direction of the vertical axis (the W axis) is the one pointing upward from the liquid's surface. Determine which of the five straight lines, *1-5*, corresponds to an object with the highest density and which to an object with the lowest density. Is there a straight line among these five for which $\rho_{ob} = (1/2) \rho_{liq}$? The arrows on the straight lines point in the direction of motion of the object.

1.30. The dependence of the potential energy W of the interaction between two objects on the distance r separating them is shown in the figure. What will be the distances between the objects that correspond to equilibrium positions? At what distance will the equilibrium be stable? (Answer the same question for unstable equilibrium.) What segments of the curve correspond to a repulsive force and what segments, to an attractive force?

1.31. A load of mass m_2 is hanging from a string. A bullet flying horizontally hits the load. Three cases are possible here, namely, (1) the bullet pierces the load and,

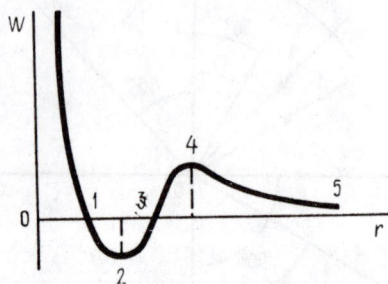

Fig. 1.30

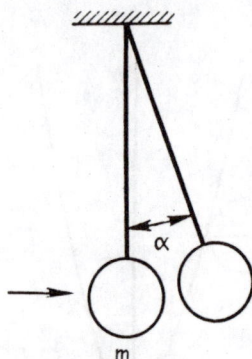

Fig. 1.31

retaining a fraction of its velocity, continues its flight, (2) the bullet gets stuck in the load, and (3) the bullet recoils from the load. In which of these three cases will the load be deflected by an angle α with the greatest magnitude and in which will it be deflected by an angle with the smallest magnitude?

1.32. Two spheres of equal mass collide, with the collision being absolutely elastic but not central. Prove that in this case the angle between the velocities after collision must be $90°$.

1.33. A sphere of mass m_1 impinges with a velocity $\mathbf{v}_0$ on a sphere of mass m_2 that is at rest, with $m_1 > m_2$. The collision is absolutely elastic but not central. By what maximal angle θ will the impinging sphere be deflected?

1.34. Two spheres of equal mass are moving at right angles with velocities that are equal in magnitude. At the moment of collision the velocity vector of sphere *1* is directed along the straight line connecting the centers of the spheres. The collision is absolutely elastic. Plot the velocity vectors before· and after collision in different coordinate systems: (1) in the laboratory system (in this system the velocities of the spheres are those specified above), (2) in the coordinate system connected with the center of mass of the two spheres, and (3) and (4) in the coordinate systems linked to each of the spheres.

1.35. The centers of the spheres *1*, *2*, and *3* lie on a single straight line. Sphere *1* is moving with an (initial) velocity v_1 directed along this line and hits sphere *2*.

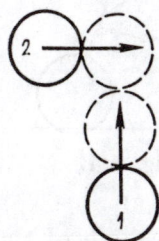

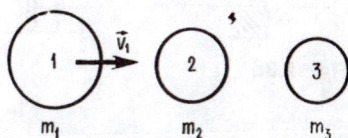

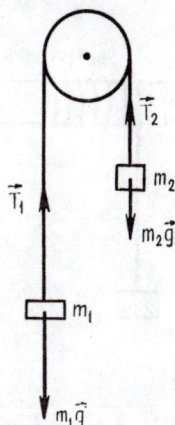

Fig. 1.34   Fig. 1.35   Fig. 1.37

Sphere *2*, acquiring after collision a velocity v_2, hits sphere *3*. Both collisions are absolutely elastic. What must be the mass of sphere *2* for the sphere *3* to acquire maximum velocity (the masses m_1 and m_3 of spheres *1* and *3* are known)?

1.36. A sphere of mass m_1 moving with a velocity v_0 hits a sphere of mass m_2 that is at rest. The collision is absolutely elastic and central. The velocities of the spheres after collision are u_1 and u_2, respectively. What are the mass ratios for the following values of velocities: $u_1 = 0$, $u_1 < 0$, and $u_1 > 0$?

1.37. A device often used to illustrate the laws of uniformly accelerated motion is the Atwood machine. The machine consists of two loads of mass m_1 and m_2 attached to the ends of a limp but inextensible string. The string runs over a pulley. The acceleration with which the loads move is

$$w = \frac{m_1 - m_2}{m_1 + m_2} g,$$

whereas the angular acceleration of the pulley is ignored. Is the last assumption true for exact calculations?

1.38. Strings are wound around a shaft and a sheave of equal mass, and a load is attached to the end of each string (the loads have equal mass). Which of the two loads

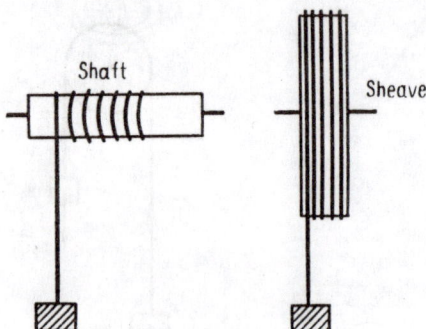

Fig. 1.38

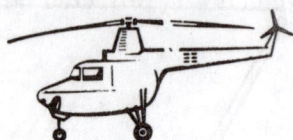

Fig. 1.40

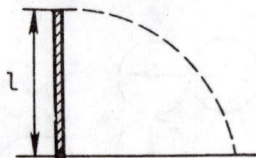

Fig. 1.41

will descend with a greater acceleration and which of the rotating objects, the shaft or the sheave, has a greater angular acceleration?

1.39. A vacuum cleaner standing on the floor turns through a small angle when switched on and then stops. Why does this happen?

1.40. A number of types of helicopters, among which are the Soviet-made "Mi" helicopters and the Westland Whirlwinds designed for use by Queen Elizabeth II, utilize one main rotor and a small vertical tail rotor. What is the function of this second rotor?

1.41. A rod whose lower end is sliding along the horizontal plane starts to topple from the vertical position. What will be the velocity of the upper end when this end hits the ground?

1.42. A thin rod of length $2R$ and mass m is standing (vertically) on a perfectly smooth floor. The state of equilibrium in which the rod is at rest is unstable, and the rod falls. Find the trajectories that the various points of the

20

rod describe and the velocity with which the upper end of the rod hits the floor.

1.43. A homogeneous rod AB is lying on a perfectly smooth floor. A bullet hits the rod and gets stuck in it. The direction of the bullet's initial velocity v_0 is perpendicular to the rod, and the point where the bullet hits the rod lies at a distance x from the middle of the rod. The mass of the bullet is m and the mass of the rod is M.

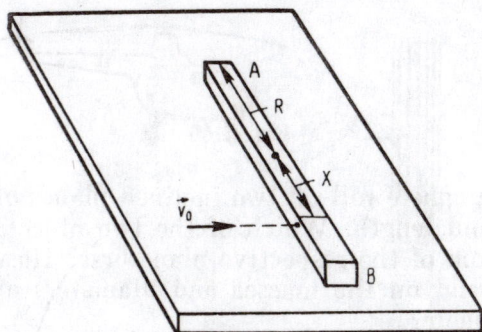

Fig. 1.43

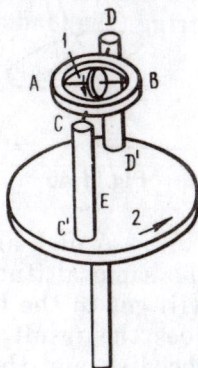

Fig. 1.44

Will a velocity directed in opposition to v_0 be imparted to end A at the first moment after the collision?

1.44. The axis AB of a gyroscope is mounted in a frame that can rotate about the axis CD. This frame is mounted, via vertical supports CC' and DD', on a horizontal platform which, in turn, can rotate about the axis EF. At first the platform is at rest and the gyroscope is rotating in the direction designated by arrow 1. Then the platform begins to rotate in the direction designated by arrow 2. How will the gyroscope's axis change its position in space?

1.45. A top is spinning in the direction designated by the arrow in the figure. In what direction does the precession of the top occur?

1.46. A shaft whose diameter is d and length is l is rotating without friction in bearings with an angular velocity ω_0. A sleeve of height h and outer diameter D is fitted on the shaft (the materials of the sleeve and the shaft are the same). At first the sleeve is not connected with the shaft and is at rest. Then at some moment the sleeve is clamped to the shaft. What will be the common angular velocity of the shaft plus the sleeve?

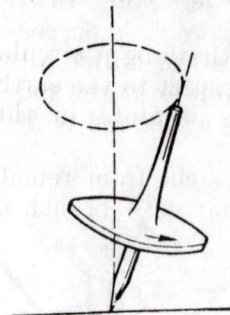

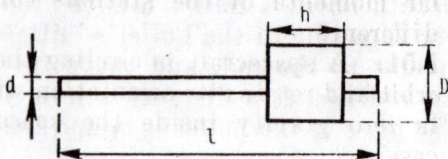

Fig. 1.45   **Fig. 1.46**

1.47. A disk and a sphere roll off two inclined planes of the same altitude and length. Which of the two objects will get to the bottom of the respective plane first? How does the result depend on the masses and diameters of the disk and the sphere?

1.48. A spacecraft is circling the earth E along an elliptical orbit. How must the velocity of the spacecraft at

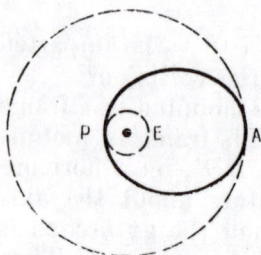

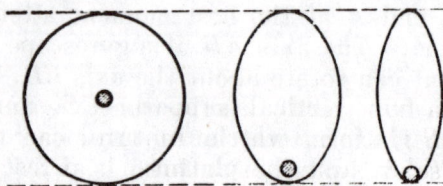

Fig. 1.48   **Fig. 1.50**

perigee P and apogee A be changed so that the spacecraft follow a circular orbit?

1.49. Several artificial satellites of the same mass are circling the earth along circular orbits of different radii. How do the kinetic, potential, and total energies and angular momenta of the satellites depend on the radii of the orbits?

1.50. Three orbital space stations are circling the earth along different orbits: one along a circular orbit and the other two along elliptical orbits whose major axes are equal to the diameter of the circular orbit. The masses of the stations are the same. Will the energies and angular momenta of the stations coincide or will they be different?

1.51. A spacecraft is circling the earth along a circular orbit and retains its orientation with respect to the earth. Is zero gravity inside the spacecraft absolute in this case?

1.52. A comet flies into the solar system from remote outer space. The trajectory of the comet is a branch of

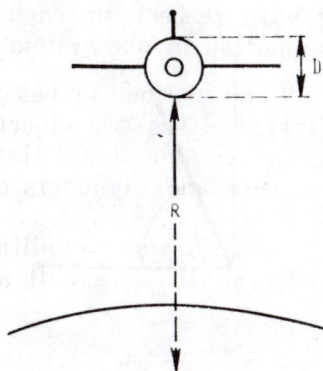

Fig. 1.52

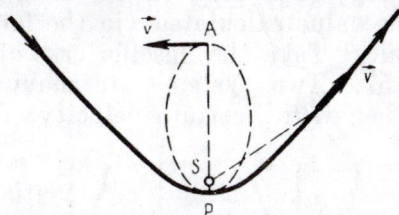

Fig. 1.51

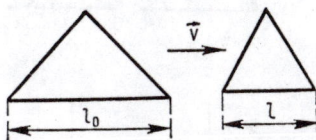

Fig. 1.54

a hyperbola. Can the comet become a satellite of the sun S if the interaction of the comet with the planets of the solar system is ignored?

1.53. What shape will a round disk have if viewed from a system of coordinates with respect to which the disk is moving with a certain velocity directed along the diameter of the disk?

1.54. An isosceles right triangle is moving with respect to a system of coordinates with a velocity **v** directed along the hypotenuse. When viewed from this system, the triangle appears to be an equilateral triangle. Find the velocity with which the triangle is moving with respect to this system.

1.55. The various relationships that exist between time intervals, coordinates, and velocities in the special theory

23

of relativity are conveniently illustrated via a system of coordinates in which on the axes we lay off either distance and time multiplied by the speed of light or time and distance divided by the speed of light. Curves that represent motion in such systems are known as world lines. Various world lines are shown in the figure in the x/c vs. t coordinates. What does each line represent? Is there a line that contradicts the main principles of relativity theory?

1.56. A world line is directed at an angle θ to the x/c axis (see Problem 1.55). What is the ratio of the kinetic energy calculated via the formula of relativity theory to the value calculated via the formula of classical mechanics? Take the specific case of $\theta = 60°$ as an example.

1.57. Two systems are moving with respect to each other with a certain velocity. The motion of one system

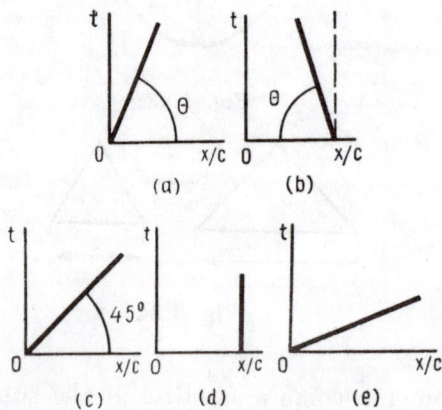

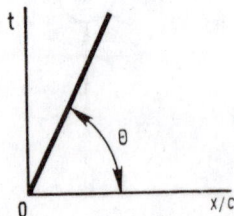

Fig. 1.55   Fig. 1.57

in terms of the coordinates x/c and t of the other system is represented by a world line directed at an angle θ to the x/c axis. After a time interval T_0 reckoned from the origin of coordinates has elapsed, one system sends a signal to the other. After what time will the second system receive the signal?

1.58. Three systems, A, B and C, are moving with respect to each other in such a manner that with respect to system B the velocities of A and C coincide in magnitude and are directed toward B (Figure (a)). When system A comes alongside system B (Figure (b)), the clocks in the

24

two systems are synchronized. At this moment system A begins emitting signals directed at B and separated by equal time intervals T_0. This continues until A comes alongside C (Figure (c)), with N signals being set over

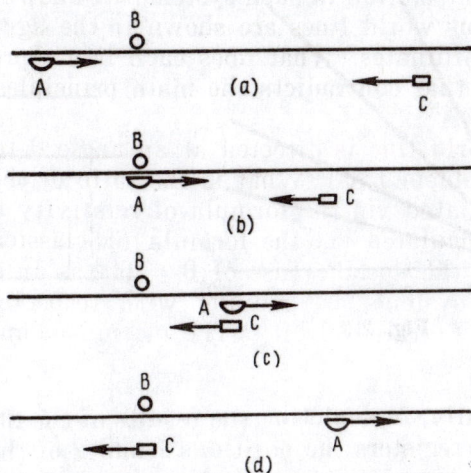

Fig. 1.58

the entire interval between the encounters. At this moment the clock in C is synchronized with the clock in A and system C starts to send signals directed at B that are separated by the same time intervals T_0. Find the difference in readings of the clock in B and C when these two systems come alongside (Figure (d)).

2. Molecular Physics and Thermodynamics

2.1. Two balloons of the same volume are filled with gases at the same pressure, one with hydrogen and the other with helium. Which of the two has the greater buoyancy (including the weight of the bag) and what is the ratio of buoyancies?

2.2. Which of the lines in the figure reflects correctly on the log-log scale the temperature dependence of the root-mean-square velocity of molecules?

2.3. Why is the trace of the silver molecules in the Stern experiment for measuring the velocities of molecules sharp in the case of fixed cylinders (Figure (a)) and blurred in the case of rotating cylinders (Figure (b))?

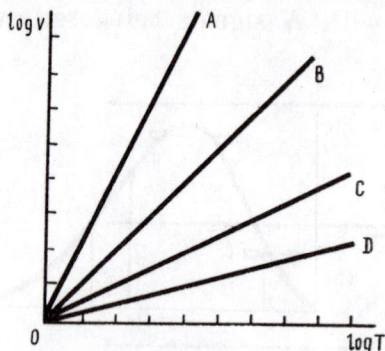

Fig. 2.2

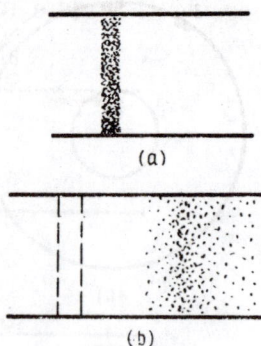

Fig. 2.3

2.4. Usually, in depicting the results of the Stern experiment, one registers the positions *1* and *2* of the traces of silver for, respectively, fixed and rotating cylinders (Figure (a)). However, a student depicted the traces in a manner shown in Figure (b). The instructor remarked that such a position of traces contradicts the experimental results, and yet the student was able to defend his position. Under what condition can such an experimental situation occur? What are the chances of encountering it in actual experiments?

2.5. The functions $F(v) = dN/dv$ and $f(v) = (1/N_0) dN/dv$, with N the number of molecules having velocity v and N_0 the total number of molecules in a given volume, are laid off on the vertical axes in Figures (a) and (b), respectively. What is the physical meaning of each hatched segment in these figures?

2.6. All the ordinates of curve *2* are twice the corresponding ordinates of curve *1*. What is the difference between the velocity distribution functions represented by these curves?

2.7. A segment from velocity v_2 to velocity v_3 on the graph representing the velocity distribution function is isolated (see Figure (b) accompanying Problem 2.5). How can we on the basis of this graph determine the energy of all the molecules whose velocities lie within the specified range and the average energy of these molecules?

2.8. The velocity distribution for molecules can be represented as a function of the ratio of the given velocity

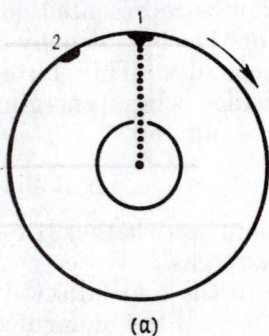

(a)

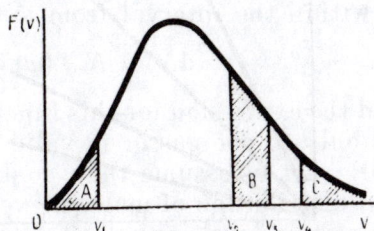

(a)

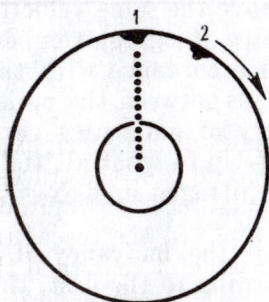

(b)

Fig. 2.4

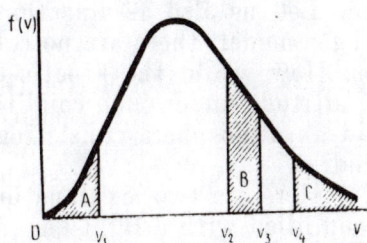

(b)

Fig. 2.5

to the most probable one. It is then expedient to lay off on the vertical axis the ratio of the value of the function for the given velocity to the value of the function for the

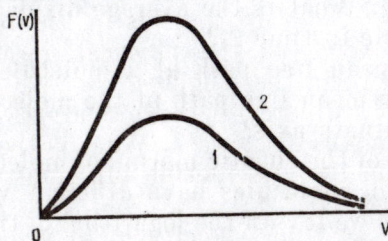

Fig. 2.6

most probable velocity. Will the distribution curve constructed in this manner be valid for different gases, different number of molecules, and different temperatures or will it be necessary to reconstruct the curve anew for each case?

2.9. The Maxwellian distribution can be represented not only by a function of molecule velocities but also by a function of the energies of the molecules. This latter function gives the number of molecules whose energies lie within the interval from w to $w + dw$, or

$$dN = N_0 f(w) \, dw. \qquad (2.9.1)$$

Find the expression for this function and see whether it refers only to one gas or is valid for any gas.

2.10. Let us assume that, contrary to the real (Maxwellian) distribution of molecule velocities, all the molecules at a certain level, say at sea level, have the same velocity equal to the root-mean-square velocity at a given temperature. Let us also assume that, in accordance with the ideal gas model, there are no collisions between the molecules. How would the kinetic energy of molecules vary with altitude under such conditions? Up to what altitude would an atmosphere consisting of nitrogen and oxygen extend?

2.11. Here are two explanations of the buoyancy of a balloon filled with a light gas. According to the first, the buoyancy is simply the Archimedes' force equal to the weight of the air that would occupy the volume of the balloon (filled with the gas), while according to the second, the buoyancy is the difference between the barometric pressures acting on the upper and lower sections of the balloon. Do these explanations contradict each other?

2.12. The average displacement of a Brownian particle in time t is $\langle l \rangle$. What is the average displacement $\langle l \rangle$ of the same particle in time $2t$?

2.13. If the mean free path of a molecule in a gas is $\langle l \rangle$, what is the mean free path of the molecule along an arbitrary coordinate axis?

2.14. Because of the chaotic motion of molecules in a gas the free paths of molecules have different values. If on the vertical axis we lay off the logarithm of the number of molecules whose free paths exceed a certain value x and on the horizontal axis the value of x, the graph representing the dependence of these two quantities is a straight line with a negative slope,

$$\log N = \log N_0 - ax.$$

How can one find the free path of molecules using such a graph?

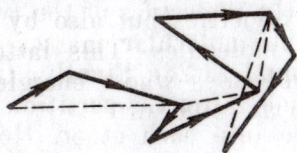

Fig. 2.12

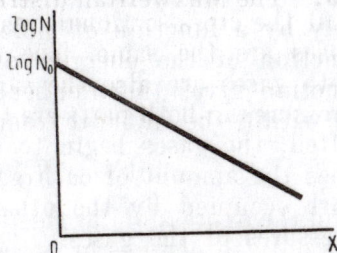

Fig. 2.14

2.15. A vessel is divided by a porous partition into two parts, 1 and 2, of equal volume. After the air was pumped out of the vessel, part 1 was filled with hydrogen and part 2 with nitrogen. The initial pressures of the gases are the same. Draw a rough sketch of the graph of how the pressures of the gases in the vessel change with the passage of time.

2.16. The temperature of a gas in a vessel changes depending on whether the vessel is open or closed, and so

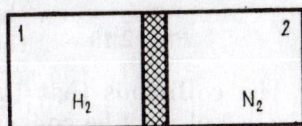

Fig. 2.15

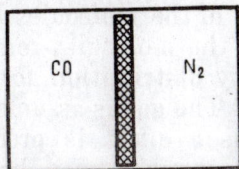

Fig. 2.16

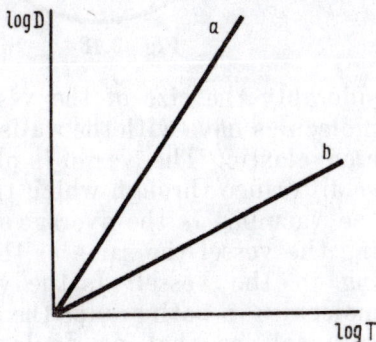

Fig. 2.17

does the diffusion coefficient. The temperature dependence of the diffusion coefficient D for both cases is shown in the figure on the log-log scale. Which line corresponds to the case of an open vessel and which to the case of a closed vessel? The effective cross sections of the molecules are assumed to be constant.

2.17. A vessel is divided by a solid partition into two parts of equal volume. One part is filled with nitrogen and the other with carbon monoxide. It may be assumed that the cross-sectional areas of the molecules of the two gases are the same. The relative molecular masses of both gases are also the same (equal to 28). Finally, the pressures in both parts are the same. After the partition is lifted, the gases begin to diffuse into each other. How does the amount of each gas that has transferred to the part occupied by the other gas depend on the initial pressures of the gases?

2.18. A gas is inclosed in a vessel and has a pressure at which the mean free path of the molecules exceeds con-

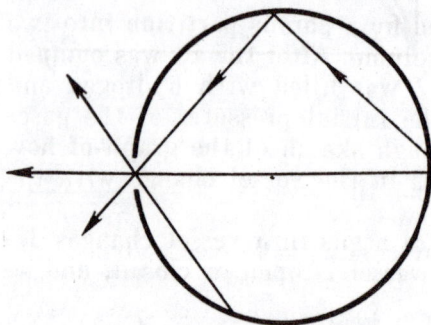
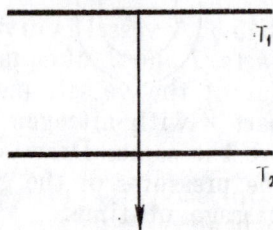

Fig. 2.18   Fig. 2.19

siderably the size of the vessel. The collisions that the molecules have with the walls of the vessel may be considered elastic. The vessel is placed in a vacuum and has a small orifice through which the gas molecules escape into the vacuum. Is the average energy of the molecules leaving the vessel the same as that of the molecules remaining in the vessel? Is the velocity distribution for the molecules in both groups the same? The gas is assumed to be ideal, so that no Joule-Thomson effect is present.

2.19. A heat flux passes through a gas from a heated plate with a temperature T_1 to a cold plate with a temperature T_2. The linear dimensions of the plates are large compared to the distance between them. Is the temperature gradient the same along the entire heat flux? Why when measuring the thermal conductivity coefficient must we place the plates horizontally, with the plate with the higher temperature placed above the one with the lower temperature?

2.20. Liquid nitrogen ($t = -196$ °C) is inside a Dewar vessel. The air surrounding the vessel has a temperature $t = 20$ °C. The pressure of the residual gas between the walls of the vessel is about 10^{-4} Pa (roughly 10^{-6} mm Hg). The mean free path of the "molecules" of air at atmospheric pressure is about 10^{-7} m. What is the temperature of the air between the walls of the vessel?

2.21. Steady-state heat transfer through a gas occurs between two parallel walls. The experiment is conduct-

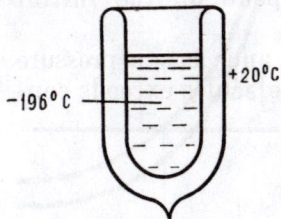

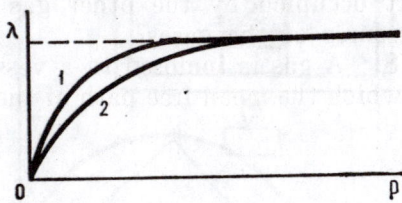

Fig. 2.20   Fig. 2.21

ed in such conditions that the only process by which the heat is transferred is pure thermal conduction. The dependence of the thermal conductivity coefficient λ is measured as a function of the gas pressure p, with the experiment conducted twice, for two different distances between the walls. The results are shown in the figure. What curve corresponds to the greater distance between the walls?

2.22. Figures (a), (b) and (c) depict three cyclic processes in the pV-, VT-, and pT-coordinates. The curvilinear

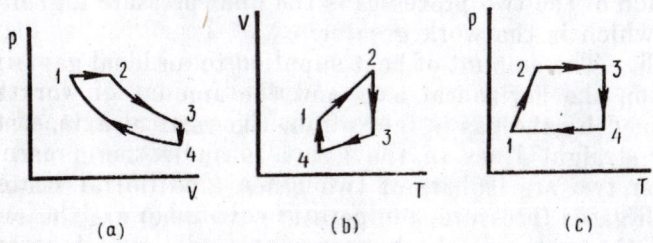

Fig. 2.22

sections in Figure (a) are isotherms. Depict the same processes in the pT- and VT-coordinates (for process (a)), the pV- and pT-coordinates (for process (b)), and the pV- and VT-coordinates (for process (c)).

2.23. A gas is inside a cylinder clo ed by a piston. The piston is held from above by a spring whose elastic properties obey Hooke's law. Produce a rough sketch, in the pV-coordinates, of the curve that represents the change in state of the gas upon heating and determine the work

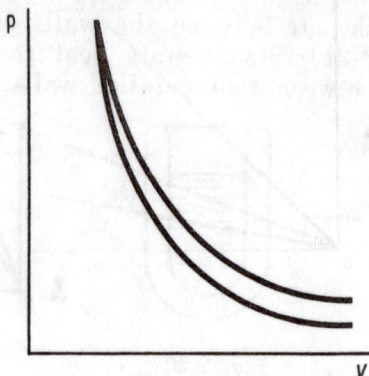

Fig. 2.23   Fig. 2.24

that is done in the process if the volume of the gas varies from V_1 to V_2 and the pressure varies from p_1 to p_2.

2.24. The figure demonstrates the adiabatic curves for two gases, helium and carbon dioxide. Which curve corresponds to which gas?

2.25. A gas expands from an initial state characterized by a pressure p_1 and a volume V_1 in two ways, isothermically and adiabatically, to the same volume V_2. In which of the two processes is the final pressure higher and in which is the work greater?

2.26. The amount of heat supplied to an ideal gas is laid off on the horizontal axis and the amount of work performed by the gas is laid off on the vertical axis. One of the straight lines in the figure is an isotherm and the other two are isobars of two gases. The initial states of both gases (pressure, temperature, volume) are the same, and the scales on the two axes coincide. Which straight line corresponds to which process? How many degrees of freedom does each gas have? (Vibrational degrees of freedom are not to be taken into account.) The graphs of what processes coincide with the coordinate axes?

2.27. The straight lines in the figure depict the variations in temperature as a function of the amount of heat supplied in different processes involving the change of state of a monatomic and a diatomic gas. Which processes correspond to these straight lines? The graphs of what processes coincide with the coordinate axes? The initial

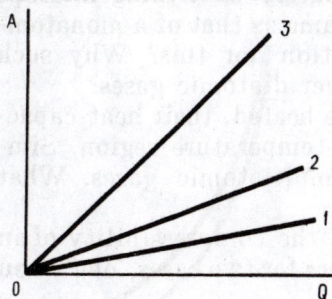

Fig. 2.26

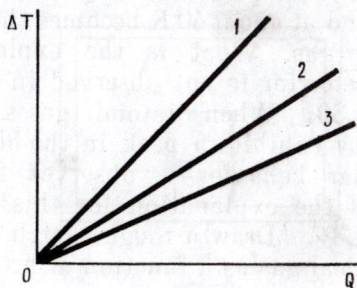

Fig. 2.27

states (temperature, volume, pressure) of the two gases are the same.

2.28. One of the straight lines in the figure depicts the dependence of the work done on the temperature variations for an isobaric process. The other two are the adiabatic curves for argon and nitrogen. Which straight line

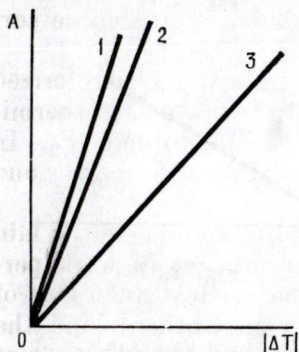

Fig. 2.28

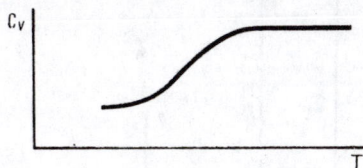

Fig. 2.29

corresponds to which process? How should one depict an isotherm and an isochor in these coordinates? Bear in mind that on the horizontal axis we lay off the difference between the higher and the lower temperature.

2.29. For temperatures close to room temperature and somewhat higher, the molar heat capacity of hydrogen agrees, with good accuracy, with the results predicted by the classical theory of heat capacity for ideal gases, a theory that allows for three translational and two rotational degrees of freedom for diatomic gases. However, at low temperatures the heat capacity of hydrogen drops and at about 40 K becomes the same as that of a monatomic gas. What is the explanation for this? Why such behavior is not observed in other diatomic gases?

2.30. When diatomic gases are heated, their heat capacity exhibits a peak in the high-temperature region. Similar behavior is observed in multiatomic gases. What is the explanation for this?

2.31. Draw a rough sketch for the compressibility of an ideal gas as a function of pressure for two cases, one when

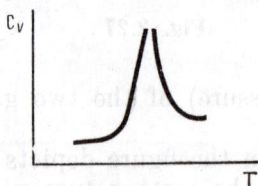

Fig. 2.30

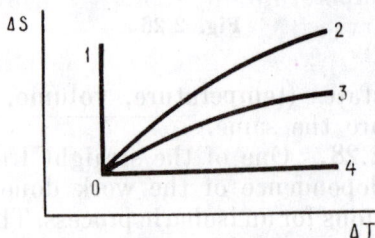

Fig. 2.34

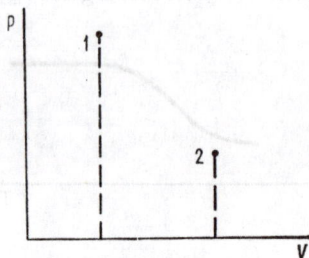

Fig. 2.32

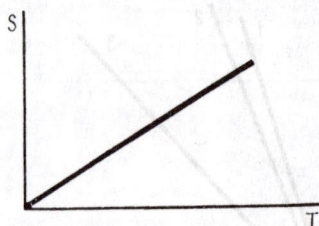

Fig. 2.37

the gas is compressed isothermically and the other when the gas is compressed adiabatically.

2.32. A gas is transferred from a state *1* to a state *2* by two processes: (a) first by an isochor and then by an isobar, and (b) first by an isobar and then by an isochor. Will the work done in both cases be the same, will the amount of heat required in the processes be the same, and will the increment of entropy in the processes be the same?

2.33. Draw the Carnot cycle for a monatomic gas on the log-log scale using the pT- and VT-coordinates.

2.34. A gas is transferred from an initial state 0 to other states 1, 2, 3, and 4 via different isoprocesses. Which curve representing the dependence of entropy on temperature corresponds to which process?

2.35. Draw the Carnot cycle in the ST-coordinates.

2.36. Two objects with initial temperatures T_1 and T_2 (with $T_1 > T_2$) are brought into contact. The objects are isolated from their surroundings, and the masses and heat capacities of the two objects coincide. How does the total entropy of these objects change as the temperatures become equal?

2.37. Suppose that the entropy grows linearly with temperature in a process. How does the heat capacity vary with temperature?

2.38. A gas is transferred from a state 1 to a state 2 in two ways: (a) directly by an isobar, and (2) first by the

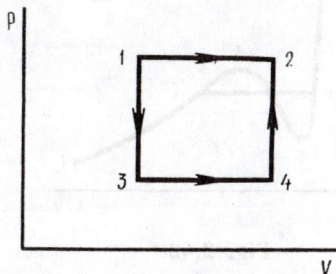

Fig. 2.38

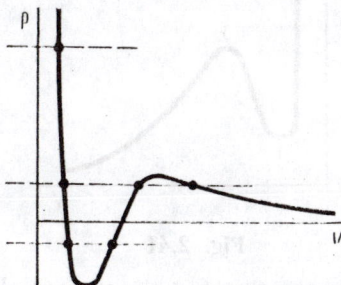

Fig. 2.40

isochor 1-3, then by the isobar 3-4, and, finally, by the isochor 4-2. Show, by direct calculation, that the entropy increment in both cases is the same.

2.39. A heat engine operates according to a cycle that consists of two isochors and two isobars. Prove that the entropy of the heater-gas-cooler system increases as the engine operates. How does the entropy of the gas change in the process? The heat capacities of the heater and cooler are assumed to be infinite.

2.40. According to the van der Waals equation, which is a third-degree equation in the volume, the theoretical isotherm of a real gas may have either one or three intersections with a horizontal line, the intersections corresponding to either one or three real roots of the equation. With three roots it may so happen that two are equal (maxima and minima on the isotherm) or even all three are equal (the critical point). However, on an isotherm built for a sufficiently low temperature there is a section lying below the horizontal axis, and a horizontal line in this case intersects the section only at two points (two roots in V). Where in this case is the compulsory third root?

2.41. The section *1-3* on the theoretical isotherm of a real gas (the van der Waals isotherm) is assumed to be unrealistic because of its absolute instability. What is the reason for this instability?

2.42. Changes in the state of a real gas or liquid that are realized under ordinary conditions at a constant

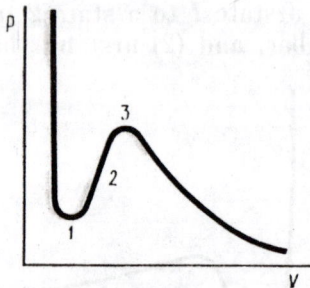

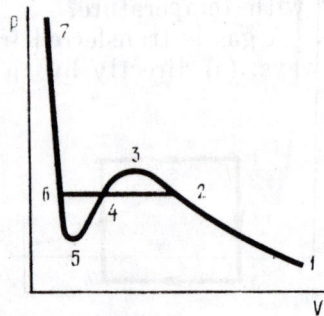

Fig. 2.41   Fig. 2.42

temperature are represented by the so-called Andrews isotherm, which consists of a section (*1-2*) representing the unsaturated vapor, a section (*2-4-6*) representing the two-phase state (saturated vapor and liquid), and a section (*6-7*) representing the liquid. This isotherm differs from the theoretical van der Waals isotherm (*1-2-3-4-5-6-7*), which corresponds to a one-phase transition of the entire mass of vapor into liquid. On the van der Waals isotherm there are sections corresponding to metastable states (*2-3* and *5-6*), which can be realized in certain conditions. What are these states and what are the conditions for their realization?

2.43. Using the second law of thermodynamics, prove that the areas of the hatched sections between the theoretical and experimental isotherms of a real gas must be equal.

2.44. When a liquid evaporates, the heat supplied to it is used partially to do work in overcoming the forces of cohesion between the molecules (the internal heat of vaporization) and partially to do work against the forces caused by external pressure (the external heat of vaporization). How to determine the external heat of vaporization from the graph representing the experimental isotherm of a real gas?

2.45. Gas cylinders and pipes intended for operation under high pressures are usually tested not by pumping

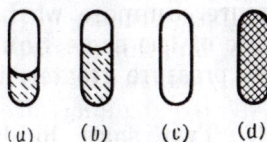

(a)   (b)   (c)   (d)

Fig. 2.46

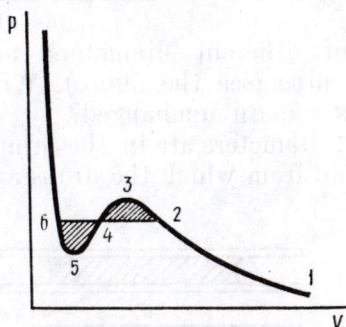

Fig. 2.43

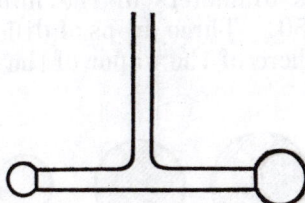

Fig. 2.49

air or a gas into them but by filling them with a liquid, water or oil, and raising the pressure up to the test value. This is done in accordance with safety regulations. What is the explanation for this?

2.46. To demonstrate the transition to the critical state, a liquid (usually ethyl ether) is placed inside a small sealed thick-walled glass tube. The tube is then sealed off (Figure (a)) and slowly heated. It is found that in the process of heating the boundary between the liquid and the vapor above the liquid rises and the meniscus becomes flatter (Figure (b)). It is extremely difficult to observe the transition through the critical temperature because of intense convective fluxes, but the result is seen because at this temperature the meniscus disappears completely (Figure (c)). Upon slowly cooling the tube it is found that at the same temperature the entire volume becomes cloudy, so that light cannot pass through the tube (Figure (d)). If the temperature is lowered still further, the volume becomes transparent and there appears a meniscus, which separates the two phases. Explain the reasons for the observed phenomena.

2.47. How does the temperature of a liquid change under adiabatic evaporation?

2.48. The bending of the surface of a liquid creates excess pressure (known as the Laplace pressure). Because of this the pressure inside a soap bubble is somewhat higher than the atmospheric pressure. In a drop, too, there is excess pressure. Suppose we have a drop of liquid and a soap bubble of the same liquid and the same diameter. Where is the pressure greater: inside the drop or inside the bubble?

2.49. Two soap bubbles of different diameters are blown out using a T-shaped pipe (see the figure). Will the diameters of the bubbles remain unchanged?

2.50. Three drops of different diameters are in the atmosphere of the vapor of the liquid from which the drops are

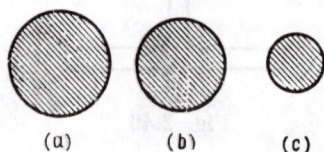

(a)　(b)　(c)

Fig. 2.50

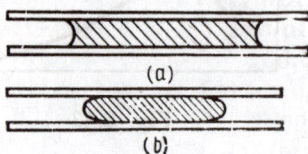

(a)

(b)

Fig. 2.51

formed. The pressure of the vapor is such that the drop with the medium diameter (Figure (b)) is in equilibrium with the vapor. Is this equilibrium stable? How will the drops of the smaller (Figure (c)) and the larger (Figure (a)) diameters behave?

2.51. Two drops are placed between two parallel glass plates, a drop of water (Figure (a)) and a drop of mercury (Figure (b)). What forces act on the plates in each case?

2.52. Inside two conical pipes there is a drop of water (Figure (a)) and a drop of mercury (Figure (b)). Where does each drop tend to move?

2.53. Which of the curves shown in the figure depicts correctly the temperature dependence of surface tension?

Curve *1* falls off to zero at the boiling point of the liquid, curve *2* falls off to zero at the critical temperature, curve *3* tends to zero asymptotically, and curve *4* shows that surface tension is temperature independent.

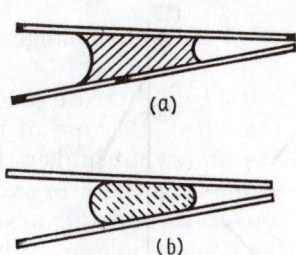

Fig. 2.52

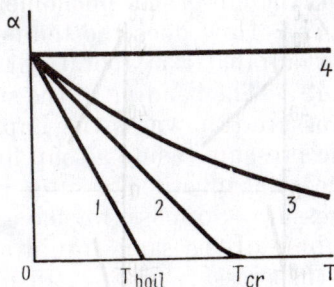

Fig. 2.53

2.54. A capillary tube is placed vertically in water. The diameter of the tube is such that surface tension "lifts" the liquid to an altitude h_0. But the height of the tube above the liquid, h, is less than h_0. How in this case will the column of liquid in the tube behave?

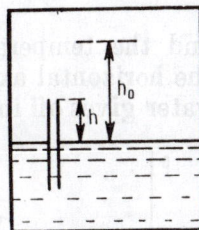

Fig. 2.54

2.55. A viscous liquid is flowing due to a pressure head Δp along a pipe of length l and diameter D. Will the volume flow remain the same if instead of this pipe we use four parallel pipes of the same length but with the diameter of each pipe being equal to $D/2$?

2.56. A viscous liquid is flowing along a horizontal pipe of diameter $D = 2R$. At some point in time a particle of rust or boiler scale gets detached from the upper part of the pipe and falls downward. Assuming that this particle acquires a constant fall velocity v_y practically at once (at this velocity the force of gravity, Archimedes' force, and the drag of the liquid balance each other), find the trajectory of the particle and the distance the particle travels in the horizontal direction due to the flow of the liquid. The maximal velocity of the liquid (along the pipe's axis) is v_{xm}.

2.57. When ice with a temperature below 0 °C is mixed with water with a temperature above 0 °C, there are four possibilities: the ice melts and the final temperature is above 0 °C, the water freezes and the final temperature is below 0 °C, part of the ice melts and the temperature of the mixture becomes 0 °C, and part of the water freezes

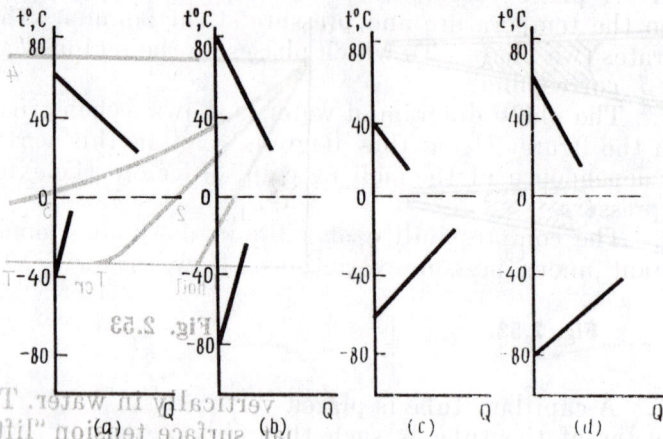

Fig. 2.57

and the temperature of the mixture becomes 0 °C. On the horizontal axis we lay off the amount of heat that the water gives off in cooling and freezing (the upper straight

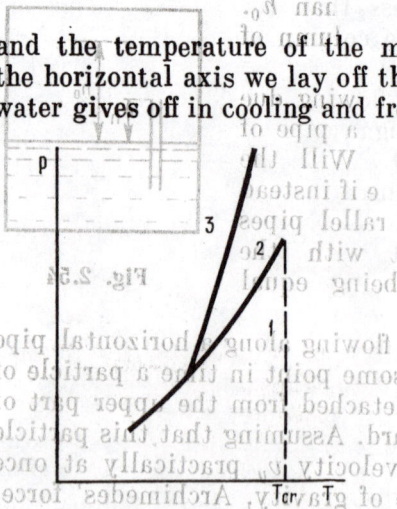

Fig. 2.58

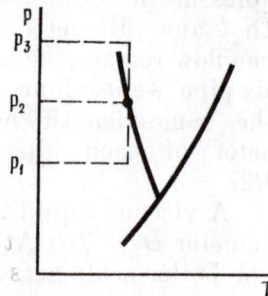

Fig. 2.59

lines) and the amount of heat that the ice absorbs in heating and melting (the lower straight lines). The scale along the horizontal axis is arbitrary, that is, the scale value is not specified. The temperature (in degrees Celsi-

40

us) is laid off on the vertical axis. Find the final result of mixing whose beginning is shown in each figure. When either all the water freezes or all the ice melts, determine the final temperature.

2.58. A phase diagram represents the relationship between the temperature and pressure at the boundary that separates two phases. To which phases do the regions *1*, *2*, and *3* correspond?

2.59. The phase diagram of water is shown schematically in the figure. Using this diagram, explain this particular dependence of the melting point of ice on the external pressure.

2.60. The compressibility of a liquid does not remain constant under pressure variations. How, knowing the

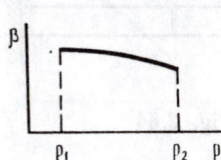

Fig. 2.60

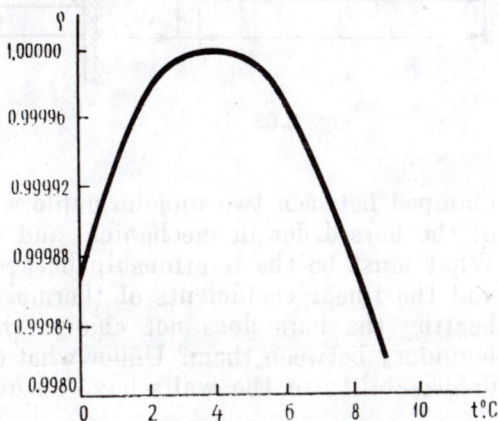

Fig. 2.61

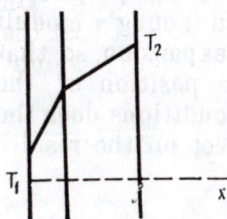

Fig. 2.62

dependence of compressibility on pressure within a certain pressure interval from p_1 to p_2, can we find the ratio of volumes at these values of pressure?

2.61. As is known, the density of water at first grows when water is heated from 0 °C but then, at 4 °C, begins to drop, as shown in the figure. Does the explanation of this lie in the fact that in introducing the metric system of units the weight of a definite volume of water at 4° C was taken as the unit of weight (subsequently this was taken as a unit of mass)?

2.62. The wall of a house consists of two layers with different thermal conductivity coefficients. The temperature of the outer wall is T_1 and that of the inner wall is T_2. Temperature variations inside the wall are shown in the figure. What layer, the inner or the outer, has a higher thermal conductivity coefficient?

2.63. A rod with a cross-sectional area S and initial length l is elongated by Δl due to a tensile stress. The modulus of longitudinal elasticity of the material of the rod, or Young's modulus, is E. Find the bulk energy density for the deformation of the rod.

2.64. Two bars *1* and *2* of the same cross-sectional area and the same length but made of different materials are

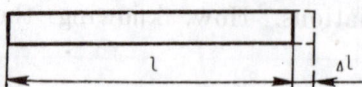

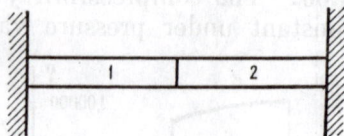

Fig. 2.63   Fig. 2.64

clamped between two undeformable walls. The materials of the bars differ in mechanical and thermal properties. What must be the relationship between Young's moduli and the linear coefficients of thermal expansion so that heating the bars does not change the position of the boundary between them? Under what conditions does the deformability of the walls have no effect on the result?

3. Electrostatics

3.1. Three charges are placed at the vertices of an isosceles right triangle, with charges $+Q$ and $-Q$ at the acute angles and a charge $+2Q$ at the right angle. Determine which of the numbered vectors coincides in direction with the field produced by these charges at a point that is the middle of the hypotenuse.

3.2. Two point-like charges *a* and *b* whose strengths are equal in absolute value are positioned at a certain distance from each other. Assuming the field strength is positive in the direction coinciding with the positive direction of the *r* axis, determine the signs of the charges for each distribution of the field strength between the charges shown in Figures (a), (b), (c), and (d).

3.3. Two point-like charges are positioned at points a and b. The field strength to the right of the charge Q_b on the line that passes through the two charges varies according to a law that is represented schematically in the figure accompanying the problem (without employing a definite scale). The field strength is assumed to be positive if its direction coincides with the positive direction on the x axis. The distance between the charges is l.

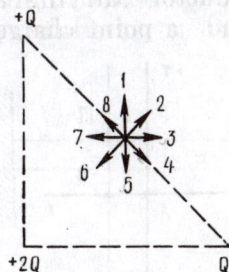

Fig. 3.1

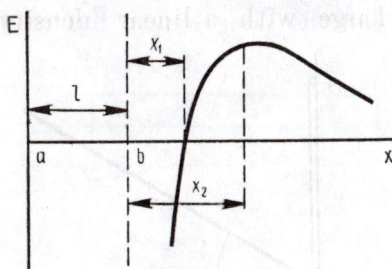

Fig. 3.3

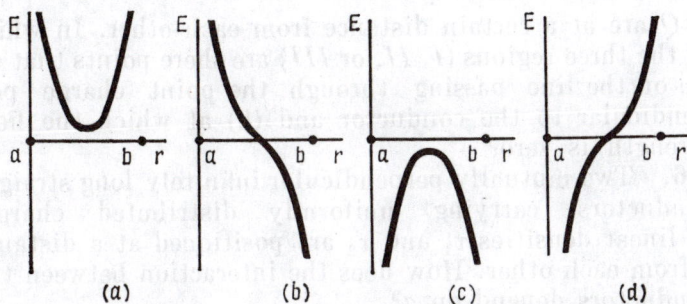

Fig. 3.2

Find the signs of the charges and, bearing in mind that the field strength at a point x_1 is zero, the ratio of the absolute values of charges Q_a and Q_b and the coordinate x_2 of the point where the field strength is maximal.

3.4. Two mutually perpendicular straight conductors carry evenly distributed charges with linear densities τ_1 and τ_2. Among the lines of force representing the field generated by these conductors there is a straight line passing through the point of intersection of the conductors.

At what angle α with respect to the conductor with the charge density τ_2 does this line pass?*

* The statement of the problem is not quite proper. The electrostatic interaction between the charges makes it impossible to maintain an even distribution of charge on the conductors. The same situation is present in other problems (e.g. see Problems 3.5 and 3.6). The difficulty can be overcome by assuming that each conductor consists of a large number of sufficiently small sections isolated from each other.

3.5. An infinitely long straight conductor carrying a charge with a linear density $+\tau$ and a point charge

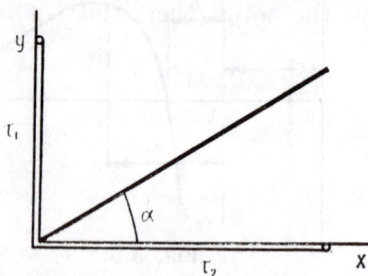 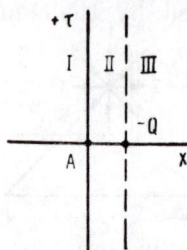

Fig. 3.4   Fig. 3.5

$-Q$ are at a certain distance from each other. In which of the three regions (I, II, or III) are there points that (a) lie on the line passing through the point charge perpendicular to the conductor and (b) at which the field strength is zero?

3.6. Two mutually perpendicular infinitely long straight conductors carrying uniformly distributed charges of linear densities τ_1 and τ_2 are positioned at a distance a from each other. How does the interaction between the conductors depend on a?

3.7. Near an infinitely large flat plate with a surface charge density σ on each side, the field strength is**

$$E = \frac{\sigma}{\varepsilon_0 \varepsilon},$$

while the field produced by a point charge at a distance r from the charge is

$$E = \frac{Q}{4\pi\varepsilon_0 \varepsilon r^2}.$$

Prove that for a uniformly charged disk with a surface charge density σ (on each side), the electric field strength on the axis of the disk is the same as for an infinitely large flat plate if the distances are small in comparison with the disk's radius R, and is the same as for a point charge if the distances are large.

** Usually the value of the field strength given in textbooks is half the one given here, since there it is assumed that the charge is on a geometric plane.

3.8. At a certain distance r from an infinitely long straight conductor with a uniformly distributed linear charge τ there is a dipole with an electric moment p_{el} directed along the line of force representing the field generated by the conductor at the point where the dipole is

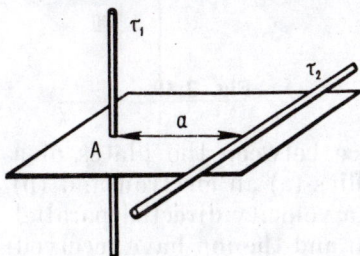

Fig. 3.6

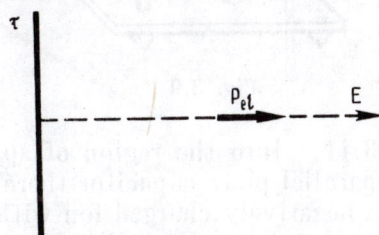

Fig. 3.8

located. Assuming the arm of the dipole is very small compared to the distance r, find the force with which the field acts on the dipole.

3.9. The figure shows the schematic of an absolute electrometer. The potential difference that is to be measured is applied between the plates *1* and *2*, with the upper plate connected to one arm of a balance beam.* The pan connected to the other arm is loaded with weights until balance is achieved, that is, when the upper plate begins to move upward. In this way the force acting between the charged plates is measured, and this enables one to determine the magnitude of the potential difference between the plates. It the equilibrium in the electrometer stable or unstable?

* The figure does not show the protecting rings around plates *1* and *2* with the same potentials. These are used to ensure that the field is as uniform as possible.

3.10. A small thin metal strip lies on the lower plate of a parallel-plate capacitor positioned horizontally. The voltage across the capacitor plates is increased gradually to a value at which the electric force acting on the strip becomes greater than the strip's weight and makes the strip move toward the upper plate. Does the force acting on the strip remain constant during the lifting process?

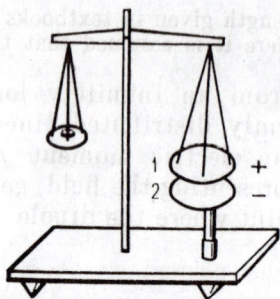

Fig. 3.9

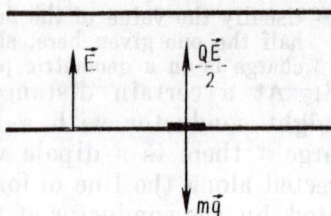

Fig. 3.10

3.11. Into the region of space between the plates of a parallel-plate capacitor there flies (a) an electron and (b) a negatively charged ion with a velocity directed parallel to the plates. Both the electron and the ion have received

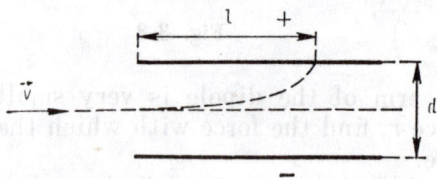

Fig. 3.11

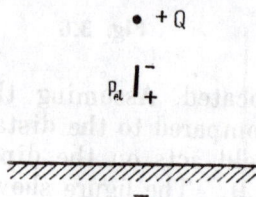

Fig. 3.12

their initial kinetic energy by passing the same potential difference U_0, and the potental difference across the capacitor is U. The distance between the plates is d. Which of the two particles will travel a greater distance before hitting the positively charged plate if both fly into the capacitor at a point that is exactly in the middle of the distance between the plates?

3.12. An electric dipole is positioned between a point-like charge and a uniformly charged conducting plate. In which direction will the dipole move?

3.13. A point-like charge Q and a dipole with an electric moment p_{el} are separated by a distance that is considerably larger than the arm of the dipole, with the result that the dipole may be considered as being point. The dipole's axis lies along the lines of force of the point charge. Compare the force acting on the dipole in the field of the point charge with that acting on the point charge in the field of the dipole.

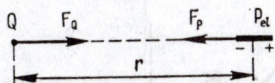

Fig. 3.13

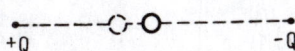

Fig. 3.14

3.14. A small uncharged sphere is positioned exactly in the midpoint between two charges whose absolute values are the same but whose signs are opposite. Suppose the sphere is shifted somewhat. Will it remain in the new position or will it move in some direction?

3.15. A small uncharged metal sphere is suspended by a long nonconducting string in the region between the

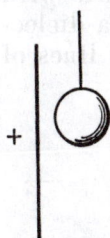

Fig. 3.15

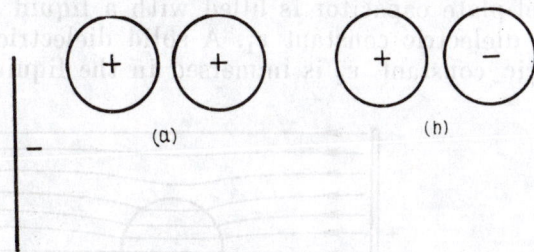

Fig. 3.16

vertically positioned plates of a parallel-plate capacitor, closer to one plate than to the other. How will the sphere behave?

3.16. Two conducting spheres carry equal charges. The distance between the spheres cannot be considered large in comparison with the diameters of the spheres. In which case will the force of interaction between the spheres be greater (in absolute value): when they carry like charges (Figure (a)) or when they carry unlike charges (Figure (b))?

3.17. A point charge is surrounded by two spherical layers (Figure (a)), with the electric field strength as a function of distance having the form depicted in Figure (b) (on the log-log scale). In what layer (the inner or the outer) is the dielectric constant greater and by what factor?

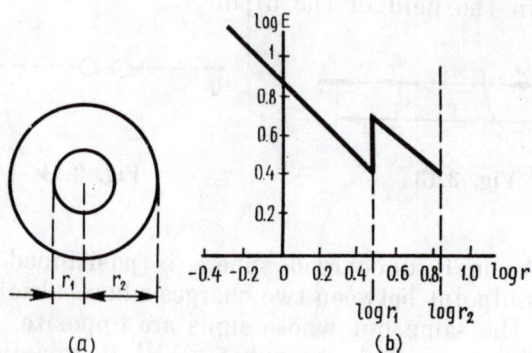

Fig. 3.17

3.18. The region of space between the plates of a parallel-plate capacitor is filled with a liquid dielectric with a dielectric constant ε_1. A solid dielectric with a dielectric constant ε_2 is immersed in the liquid. The lines of

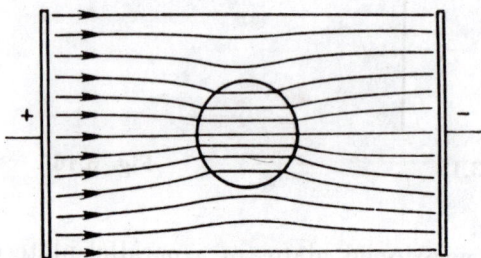

Fig. 3.18

force in the liquid have the shape shown in the figure. Which of the two dielectric constants is greater?

3.19. Various potential distributions between two point charges are shown in Figures (a)-(d) (the charges are equal in absolute value). Determine the signs of the charges for each case.

3.20. Two point charges, Q_1 and Q_2, are positioned at a certain distance from each other. The curves in the figure represent the distribution of the potential along the straight line connecting the two charges. At which points (*1*, *2*, and/or *3*) is the electric field strength zero?

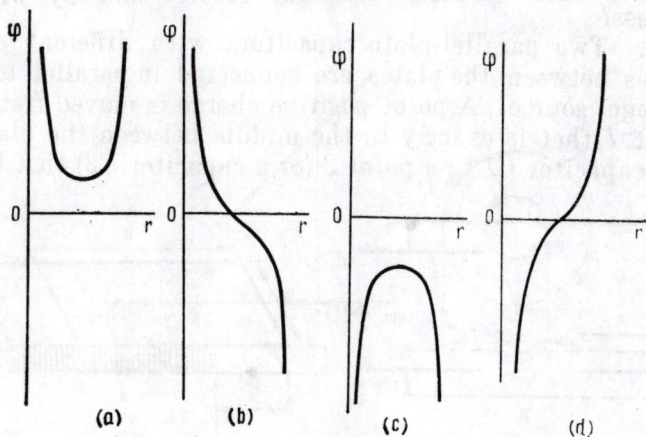

Fig. 3.19

What are the signes of the charges Q_1 and Q_2 and which of the two is greater in magnitude?

3.21. Two equal like charges are positioned at a certain distance from each other. How do the electric field

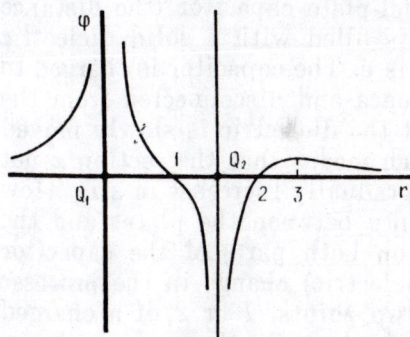

Fig. 3.20

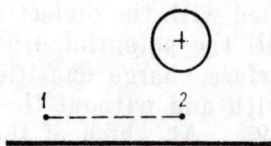

Fig. 3.22

strength and the potential vary along the axis that passes through the midpoint of the distance between the charges at right angles to the line connecting the charges?

3.22. A potential difference is applied between a conducting sphere and a conducting plate ("plus" on the sphere and "minus" on the plate). The dimensions of the plate are much larger than the distance between sphere and plate. A point positive charge is moved from point *1* to point *2* parallel to the plate. Is any work done in the process?

3.23. Two parallel-plate capacitors with different distances between the plates are connected in parallel to a voltage source. A point positive charge is moved from a point *1* that is exactly in the middle between the plates of a capacitor *C1* to a point *2* (or a capacitor *C2*) that lies

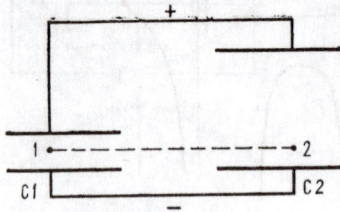
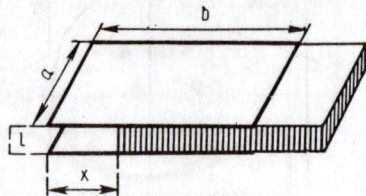

Fig. 3.23   Fig. 3.24

at a distance from the negative plate of *C2* equal to half the distance between the plates of *C1*. Is any work done in the process?

3.24. The space between the rectangular plates (with sides a and b) of a parallel-plate capacitor (the distance between the plates is l) is filled with a solid dielectric whose dielectric constant is ε. The capacitor is charged to a certain potential difference and disconnected from the voltage source. After that the dielectric is slowly moved out of the capacitor, which means that the section x not filled with the dielectric gradually increases in size. How will the potential difference between the plates and the surface charge densities on both parts of the capacitor (with and without the dielectric) change in the process?

3.25. At which of the two points, *1* or *2*, of a charged capacitor with nonparallel plates is the surface charge density greater?

3.26. The diameter of the outer conductor of a cylindrical capacitor is D_2. What should the diameter of the core, D_1, of this capacitor be so that for a given potential difference between the outer conductor and the core the electric field strength at the core is minimal?

3.27. Four capacitors, *C1*, *C2*, *C3*, and *C4*, are connected as shown in the figure. A potential difference is applied between points A and B. What should the relationship between the capacitances of the capacitors be so that the potential difference between points a and b is zero?

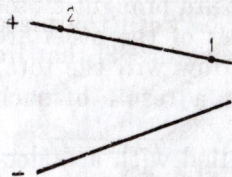

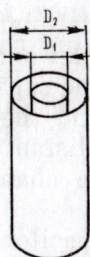

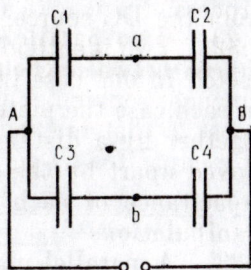

Fig. 3.25   Fig. 3.26   Fig. 3.27

3.28. An electric charge with a constant volume density ρ is distributed within a solid sphere of radius R. Determine and represent graphically the radial distributions of the electric field strength and the potential inside and outside the sphere.

3.29. In the region of space between the plates of a parallel-plate capacitor there is a uniformly distributed positive charge with a volume density ρ. The plates are connected electrically and their potential is set at zero. Calculate and draw a sketch of the distributions of the potential and electric field strength between the plates.

3.30. Two series-connected capacitors of the same size, one filled with air and the other with a dielectric, are

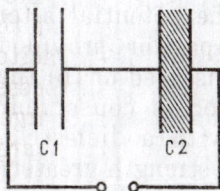

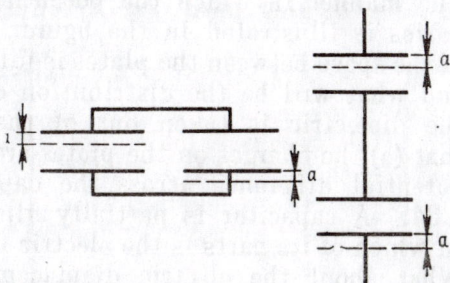

Fig. 3.30   Fig. 3.32

connected to a voltage source. To which of the capacitors a higher voltage is applied?

3.31. Two identical air capacitors are connected in series. How will the charge on and potential difference across each capacitor change when the distance between the plates of one capacitor is increased in the following cases: when the capacitors are connected to a DC source, and when the capacitors are first charged and then disconnected from the DC source?

3.32. Two identical parallel-plate air capacitors are connected in one case in parallel and in the other in series. In each case the plates of one capacitor are brought closer together by a distance a and the plates of the other are moved apart by the same distance a. How will the total capacitance of each system change as a result of such manipulations?

3.33. A parallel-plate capacitor is filled with a dielectric up to one-half of the distance between the plates.

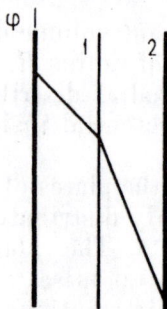   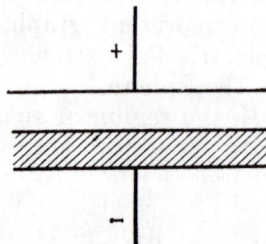

Fig. 3.33   Fig. 3.34

The manner in which the potential between the plates varies is illustrated in the figure. Which half (*1* or *2*) of the space between the plates is filled with the dielectric and what will be the distribution of the potential after the dielectric is taken out of the capacitor provided that (a) the charges on the plates are conserved or (b) the potential difference across the capacitor is conserved?

3.34. A capacitor is partially filled with a dielectric. In which of its parts is the electric field strength greater? What about the electric displacement and the energy density?

3.35. Two parallel-plate capacitors, one filled with air and the other with a dielectric, have the same geometric dimensions, are connected in parallel, and are charged to a certain potential difference. In which of the two capacitors is the electric field strength greater, in which is the electric displacement greater, in which is the energy density greater, and in which is the surface charge density on the plates greater?

3.36. Three point-like charges are positioned at the vertices of an equilateral triangles. Two are equal in magnitude and are like, while the third is opposite in sign.

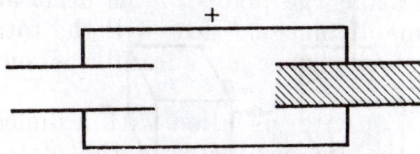

Fig. 3.35

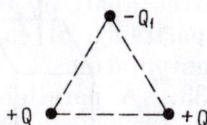

Fig. 3.36

What should the magnitude of the third charge be so that the total interaction energy of the charges is zero?

3.37. The dielectric filling the space between the plates of a capacitor that has been charged and then disconnected from the voltage source is removed. How should the distance between the plates be changed so that the energy stored in the capacitor remains the same? Explain the origin of the change in energy.

3.38. A capacitor between whose plates there is a dielectric with a dielectric constant ε is connected to a DC source. How will the energy stored in the capacitor change if the dielectric is removed? Explain the cause of this change.

3.39. A parallel-plate capacitor that has been first charged and then disconnected from the voltage source is submerged in the vertical position into a liquid dielectric. How does the level of the dielectric between the plates change in the process?

3.40. A parallel-plate capacitor with vertical plates is connected to a voltage source and then submerged into a liquid dielectric. How does the level of the dielectric between the plates change in the process? Explain the change of the energy stored by the capacitor.

3.41. A cube has been cut out from a piezoelectric crystal. When the cube was compressed, it exhibited electric charges on the faces: a positive charge on the upper face and a negative charge on the lower (Figure (a)). When the cube was stretched, the charges were found to change their signs (Figure (b)). What will be the signs of the charges on these faces if pressure is applied as shown in Figure (c)?

3.42. The relationship that exists between the electric displacement and the electric field strength in a ferroelectric is given by the curve of primary polarization and a hysteresis loop. Are there any points on the hysteresis

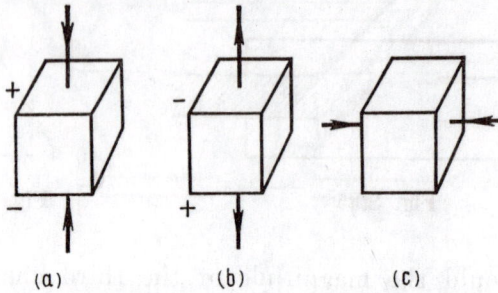

(a)   (b)   (c)

Fig. 3.41

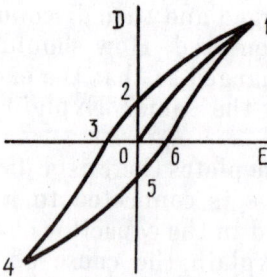

Fig. 3.42

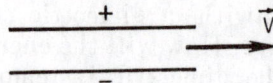

Fig. 3.43

loop to which we might formally assign a dielectric constant equal to zero or to infinity?

3.43. A charged parallel-plate capacitor is moving with respect to a certain system of coordinates with a velocity **v** directed parallel to the plates. What is the ratio of the electric field between the plates in this coordinate system to the same quantity in the system of coordinates in which the capacitor is at rest?

4. Direct Current

4.1. Two conductors, *1-3-5* and *2-4-6*, connected points with equal potentials on the resistors R_a and R_b, so that no current flows through either of them. Will there be currents flowing through them and through the 3-4 section if the key K is closed? Will this lead to a change in the reading of the ammeter?

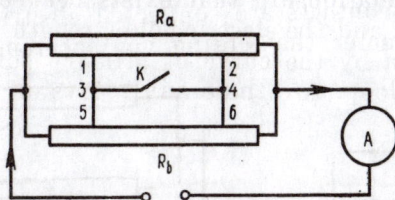

Fig. 4.1

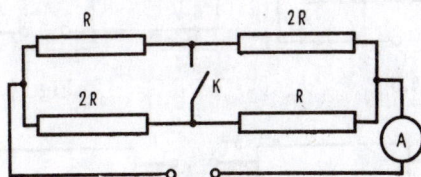

Fig. 4.2

4.2. How will the reading of the ammeter change if the key K is closed?

4.3. A voltage U_0 is applied to a potentiometer whose sliding contact is exactly in the middle. A voltmeter V is connected between the sliding contact and one fixed end of the potentiometer. It is assumed that the resistance of the voltmeter is not very high if compared with the resistance of the potentiometer. What voltage will the voltmeter show: higher than, less than, or equal to $U_0/2$?

4.4. A "black box" is an electric unit with four terminals, 1, 2, 3, and 4, and an unknown internal structure. The box shown in Figure (a) and (b) possesses the following properties: if a constant voltage of 220 V is applied to terminals 1 and 2, a voltage of 127 V appears across terminals 3 and 4 (Figure (a)), while if a voltage of 127 V is applied to terminals 3 and 4, the same voltage of 127 V appears across terminals 1 and 2 (Figure (b)). What is inside the "black box"? The formulation of the problem is quite meaningful if the voltages are measured by electrostatic voltmeters, which do not consume electric current. If voltmeters of the magnetoelectric, thermal, or electromagnetic type are employed, the voltages

across the "out" terminals of the "black box" may somewhat differ from the ones indicated in Figures (a) and (b).
4.5. Two potentiometers are connected in series, and
their sliding contacts are connected electrically, too. In
one potentiometer the sliding contact remains fixed at

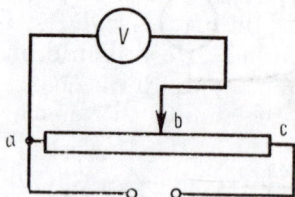

Fig. 4.3

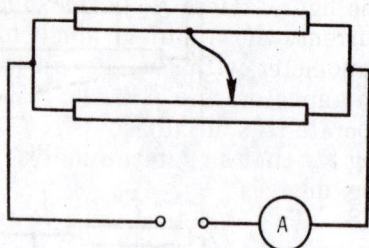

Fig. 4.5

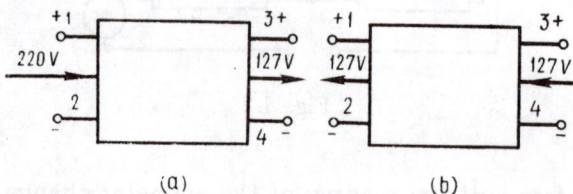

(a)   (b)

Fig. 4.4

the midpoint. How will the reading of the ammeter vary
as the sliding contact of the second potentiometer is
moved from one end of the potentiometer to the other?
4.6. A constant voltage U_0 is applied to a potentiometer of resistance R connected to an ammeter. A constant

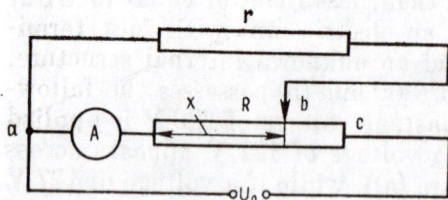

Fig. 4.6

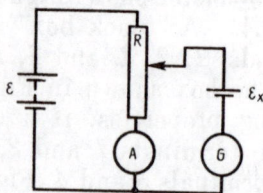

Fig. 4.7

resistor r is connected to the sliding contact of the potentiometer and the fixed end of the potentiometer (after an
ammeter). How will the reading of the ammeter vary as
the sliding contact is moved from one end of the poten-

tiometer to the other? The resistance of the ammeter is assumed to be negligible.

4.7. To measure a small emf (of, say, a galvanic cell or a thermocouple) the so-called balancing method is employed. The circuit diagram of this method is shown in the figure. Here $\mathscr{E}_x$ is the sought emf, $\mathscr{E}$ is the source of current whose emf is much higher than $\mathscr{E}_x$, G is a galvanometer with the zero in the middle of the scale, A is an ammeter, and R is the resistance box. How should one operate this circuit so as to ensure an accuracy in measuring $\mathscr{E}_x$ that is determined by the precision of the measuring devices?

4.8. Two resistors with resistances R_1 and R_2 are connected in series, and so are two capacitors with capaci-

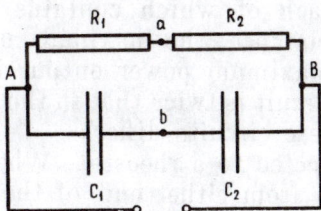

Fig. 4.8

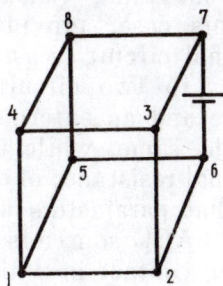

Fig. 4.10

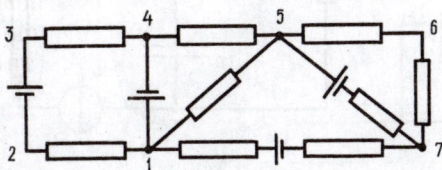

Fig. 4.9

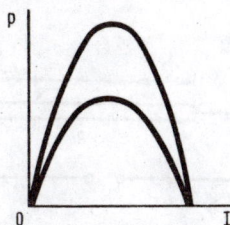

Fig. 4.12

tances C_1 and C_2. The two systems are connected in parallel and an external voltage is applied to the new system (see the figure accompanying the problem). What must be the relationship between R_1, R_2, C_1, and C_2 for the potential difference between the points a and b to be zero?

4.9. All the resistances and emf's shown in the figure accompanying the problem are assumed known. How many values of current can exist for such a circuit? How

many equations for finding these **values must we** construct on the basis of Kirchhoff's first law and **how** many
must we construct on the basis of Kirchhoff's second law?
4.10. Twelve conductors are connected in such a way
that they form a cube, and an emf source is connected
into an edge of the cube. All the resistances and the emf s
are known. There are eight junctions (eight vertices of
the cube) and six loops (six faces of the cube) in the circuit. Construct the equations for determining all the currents in the circuit.
4.11. A source of electric current with an emf $\mathcal{E}_0$ and
an internal resistance r is connected to an external circuit
with a resistance R. What must be the relationship between r and R for the power output in the external circuit
to be maximal? What is the efficiency of the current source
in this case, provided that the power output in the
external circuit is assumed to be the useful output?
4.12. In two circuits, each of which contains a DC
source and an external resistance, the maximal currents
are the same, while the maximum power output in the
external resistance of one circuit is twice that in the other.
In what parameters do these circuits differ?
4.13. ADC source is connected to a rheostat. When the
sliding contact is x distant from either end of the rheo-

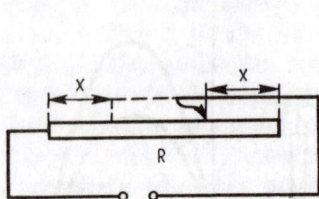

Fig. 4.13

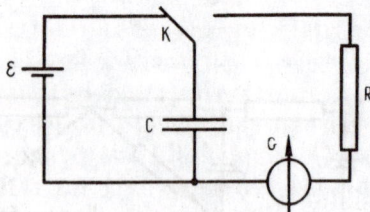

Fig. 4.16

stat (the length of the rheostat is set at unity), the power
output in the rheostat is the same in both cases. Determine the internal resistance of the DC source if the resistance of the theostat is R.
4.14. How must a large number of galvanic cells, each
having the same emf $\mathcal{E}$ and the same internal resistance
r, be connected so that in an external circuit whose resistance is R the power output is maximal?

4.15. Can a circuit be constructed in which the displacement current in the capacitor remains practically constant over a definite time interval?

4.16. A DC source with known emf $\mathscr{E}$ is charging a capacitor C. After the charging process has been completed, the capacitor is disconnected, via a key K, from the DC source and is connected to a resistor R, through which the capacitor discharges. The capacitance of the capacitor and the resistance of the resistor are selected in such a way that the charging process takes several minutes, so that the discharge current can be registered by a measuring device, G. The results of measurements are used to draw a rough curve on a diagram in which the time of discharge is laid off on the horizontal axis and the logarithm of the current, or the vertical axis. Determine the law by which the current varies and the curve representing the dependence of the logarithm of the current on the time of discharge. How can the curve help in determining the parameters of the discharge circuit, R and C?

4.17. A capacitor of capacitance C is charged to a potential difference U_0 and is then discharged through a re-

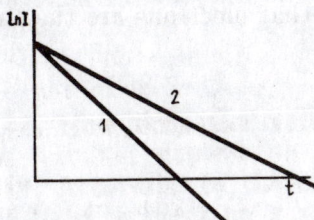
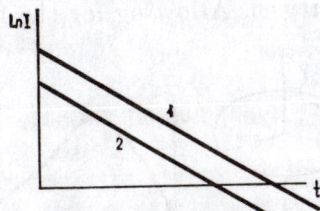

Fig. 4.17   Fig. 4.18

sistance R. The discharge current gradually decreases, with a straight line 1 corresponding to this process (see the figure accompanying the problem, where time is laid off on the horizontal axis and the logarithm of the current, on the vertical axis). Then one of the three parameters, U_0, R, or C, is changed in such a manner that the ln I vs. t dependence is represented by the straight line 2. Which of the three parameters was changed and in what direction?

4.18. A charged capacitor is discharged through a resistor two times. The time dependence of the logarithm of the discharge current obtained in the two experiments is represented by the two straight lines, 1 and 2, in the figure accompanying the problem. The experimental conditions differed only in one of the three parameters: the initial voltage of the capacitor U, the capacitance C, or the resistance R. Determine the parameter that was varied in these experiments and in which case this parameter is greater.

4.19. Prove that when a capacitor of capacitance C that has been charged to a potential difference U_0 is discharged through a resistance R, the amount of heat liberated in the conductors is equal to the initial energy stored in the capacitor.

4.20. Prove that when a capacitor is charged through a resistor R from a DC source with an emf equal to $\mathscr{E}$ half of the energy supplied by the source goes to the capacitor and half, to heating the resistor.

4.21. A charged capacitor is connected to an uncharged capacitor with the same capacitance. Determine the changes in the energies stored by the two capacitors and explain the origin of these changes from the viewpoint of energy conservation.

4.22. A conducting disk is rotating with an angular velocity ω. Allowing for the fact that electrons are the cur-

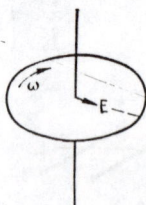

Fig. 4.22

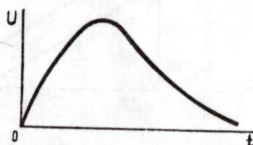

Fig. 4.23

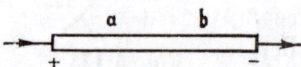

Fig. 4.24

rent carriers in a conductor, determine the potential difference between the center of the disk and the edge.

4.23. In the Tolman-Stewart experiment, a cylinder is mounted on a shaft and is rotated very rapidly. The surface of the cylinder is wound with many turns of wire of length l in a single layer. After the cylinder has been set spinning at a large angular velocity, it is braked to a stop as quickly as possible. In the circuit consisting of the wire and a measuring device, this braking manifests itself in a pulse of current caused by the potential difference that appears between the ends of the wire. If the potential difference is registered by an oscillograph, we obtain a curve similar to the one shown in the figure accompanying the problem, where time is laid off on the horizontal axis.* How, knowing the initial linear velocity of the winding, the length of the wire, and the voltage oscillogram, can one determine the electron charge-to-mass ratio?

> * In the Tolman-Stewart experiment, the quantity measured was not the potential difference but the amount of electricity passing through the circuit. This was done using a device called the ballistic galvanometer.

4.24. The section of a conductor between the points a and b is being heated. Does this lead to a redistribution of potential along the conductor (the arrow indicates the direction in which the current is flowing)? Will the passage of current change the temperature distribution in the conductor?

4.25. A constant voltage is applied to a metal wire. The current passing through the wire heats the wire to a certain temperature. Then half of the wire is cooled by a

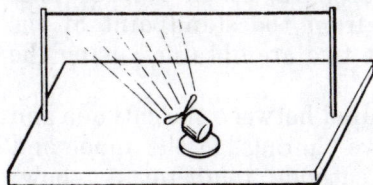

Fig. 4.25

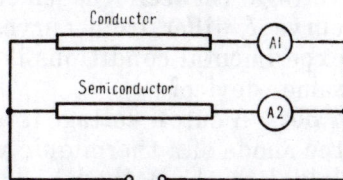

Fig. 4.27

stream of air from a fan. How will the temperature of the other half of the wire change in the process?

4.26. Two electric bulbs whose rated voltage is 127 V and whose rated wattages are 25 and 150 W are connected in series to a DC source of 220 V. Which of the two bulbs will burn out?

4.27. A conductor and a semiconductor are connected in parallel. At a certain voltage both ammeters register the same current. Will this condition remain as such if the voltage of the DC source is increased?

4.28. A conductor and a semiconductor are connected in series. The voltage applied to this system is selected in such a way that the readings of the voltmeters *V1* and *V2* coincide. Will this condition remain unchanged if the voltage of the DC source is increased?

4.29. A thermionic valve, or diode, has a heated fila-ment and a plate near it. The dependence of the current flowing between filament and plate on the voltage applied to valve (the current-voltage characteristic) is as follows. First the current grows with voltage, but then goes into a plateau at a sufficiently high voltage. Why, notwithstand-ing the fact that the filament may emit the number of

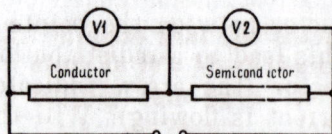

Fig. 4.28

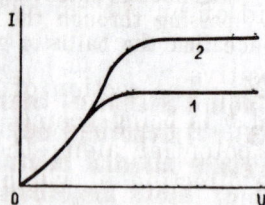

Fig. 4.29

electrons required for the saturation current to set in, the latter does not manifest itself at an arbitrarily small voltage between the electrodes? In which respect does curve *1* differ from curve *2* from the standpoint of the experimental conditions if the two are obtained using the same device?

4.30. A cutoff voltage is applied between the cathode and the anode of a thermionic valve ("minus" at the anode and "plus" at the cathode). The cathode temperature, how-ever, is sufficient for thermionic emission to manifest itself. If the direction of the electric field is reversed by applying between the cathode and the anode a voltage at which saturation current will flow through the valve, will the temperature of the cathode maintained in the cutoff di-rection of the field remain the same?

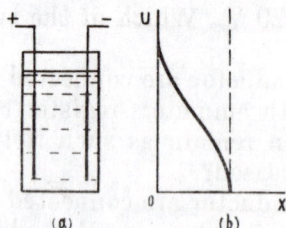

Fig. 4.31

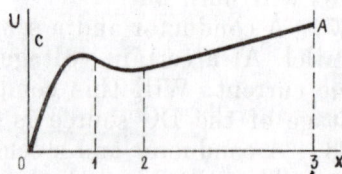

Fig. 4.32

4.31. For a current passing through an electrolyte (Fig-ure (a)), the distribution of potential between the electrodes is shown in Figure (b). Why, notwithstanding the fact that the electrodes are flat and the distance between them is much smaller than their linear dimensions, is the field between the electrodes nonuniform?

4.32. The distribution of potential between the cathode and anode in a glow discharge is shown in the figure accompanying the problem (the distance from the cathode is laid off on the horizontal axis). Within which regions of space (see the numbers on the horizontal axis) is there a positive volume charge, a negative volume charge, and a volume charge that is practically zero?

4.33. In the plasma of a gas discharge, the concentration of electrons and that of positive ions are practically the same. Does this mean that the current densities created by the motion of electrons and ions are also the same? Will an ammeter connected in series with the gas discharge gap show the sum of the electron and ion currents or their difference?

4.34. A negatively charged particle is accelerated in its motion from a cathode C to an anode A, passes through an aperture in the latter, and moves toward a Faraday cylinder F that is at the same potential as the anode (Figure (a)). For the sake of simplicity it will be assumed that the particle moves from A to F with a constant velocity. Determine the moment of time when a measuring device G in the circuit will register a current (the time is reckoned from the moment when the particle leaves the anode) and the form of the current, that is, whether the current is in the form of a pulse when the particle leaves the anode (Figure (b)) or whether it is a pulse when the particle enters the Faraday cylinder (Figure (c)) or whether there are two pulses (one when the particle leaves the anode and the other when the particle enters the Faraday cylinder; see Figure (d)) or whether the current is steady over the entire motion of the particle from the anode to the Faraday cylinder (Figure (e)).

4.35. The behavior of the potential energy of an electron inside and outside a metal is shown for two metals in Figures (a) and (b). The same figures indicate the limiting kinetic energies W_F of electrons in the metals (the Fermi levels) at $T = 0$ K. If the metals are brought into contact, what will be the values of the internal and external contact potential differences? In which metal will the electron concentration be higher?

4.36. The energy distribution function for electrons in a metal at absolute zero can be written as follows:

$$f(W) = CW^{1/2}, \qquad (4.36.1)$$

where C is a constant coefficient that is a combination of universal constants. This function terminates at W_F, which is the limiting energy, or the Fermi level. Using

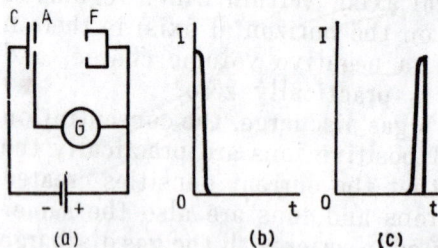

(a)   (b)   (c)

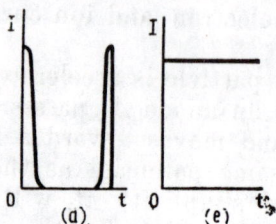

(d)   (e)

Fig. 4.34

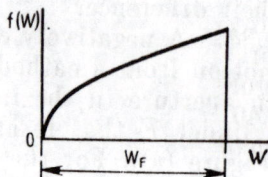

Fig. 4.36

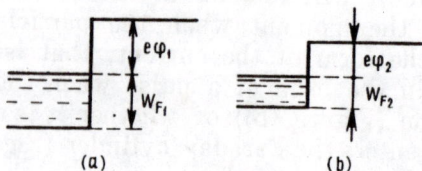

(a)   (b)

Fig. 4.35

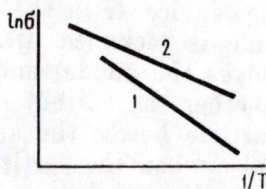

Fig. 4.37

(4.36.1), establish how the limiting energy depends on electron concentration.

4.37. The dependence of the logarithm of conductivity, $\ln \sigma$, on T^{-1}, where T is the temperature, for two semiconductors is shown in the figure. In which of the two semiconductors is the gap (the forbidden band) between the valence band and the conduction band wider?

4.38. The dependence of the logarithm of conductivity, $\ln \sigma$, on $1/T$ for two semiconductors is shown schematically in the figure. In which respect do these semiconductors differ?

4.39. The distribution of potential near the boundary between two semiconductors with different types of conduction depends on the direction of the applied external

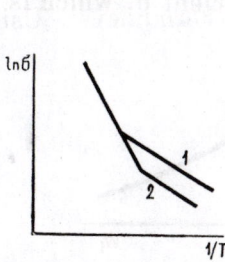

Fig. 4.38

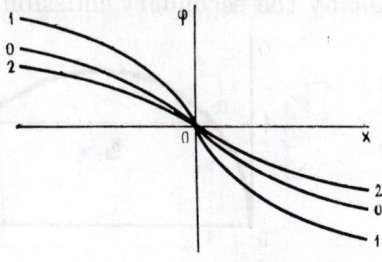

Fig. 4.39

voltage. Which distribution corresponds to the blocking direction and which, to conduction? To what semiconductors do the left and right branches of the curves in the figure belong?

4.40. The current-voltage characteristic of a semiconductor diode based on the properties of the p-n junction

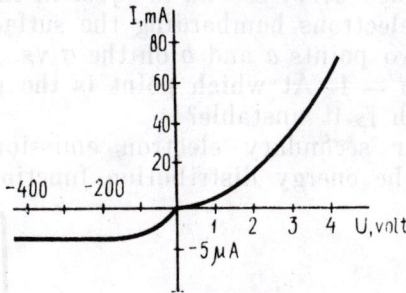

Fig. 4.40

has two branches: the upper right branch and the lower left branch. Since the right branch corresponds to small voltages and the left branch to considerably higher voltages (with the currents in the conductive direction being much higher than the currents in the blocking direction), the two branches are constructed using different scales. What is the explanation for the existence of the left branch and in what manner does the current in the blocking direction depend on the temperature of the diode?

4.41. The phenomenon of secondary electron emission consists in the following. When electrons bombard a solid surface, the surface emits secondary electrons (and partially reflects the primary electrons, which impinge on the surface). Secondary electron emission is characterized by the secondary emission coefficient σ, which is the

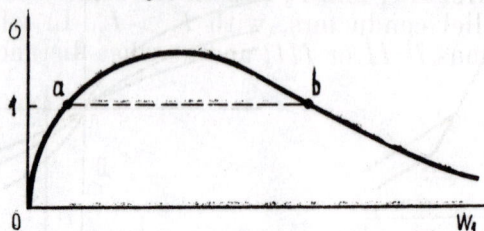

Fig. 4.41

ratio of the secondary electron current to the primary current. The dependence of the secondary emission coefficient on the primary electron energy W_1 for a certain dielectric is depicted in the figure. At σ = 1 the surface of the dielectric does not change its potential under electron bombardment, since the number of electrons leaving the surface every second is equal in this case to the number of electrons bombarding the surface every second. The two points a and b on the σ vs. W_1 curve correspond to σ = 1. At which point is the process stable and at which is it unstable?

4.42. Under secondary electron emission (see Problem 4.41), the energy distribution function $F(W_2)$ for

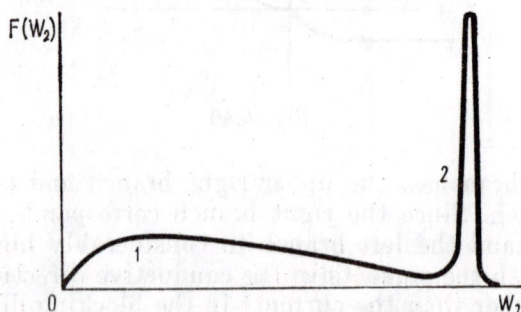

Fig. 4.42

secondary electrons is represented sufficiently well by two curves (1 and 2) shown in the figure accompanying the problem. Which of the two curves represents the primary electrons and which, the "true" secondary electrons?

5. Electromagnetism

5.1. Currents I_1 and I_2 flow in the same direction along two parallel conductors, with $I_1 > I_2$. In which of the three regions I, II or III, and at what distance from the

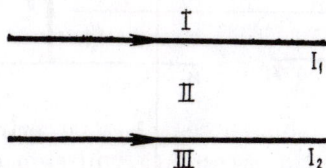

Fig. 5.1

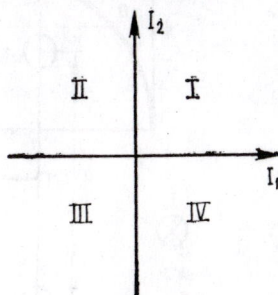

Fig. 5.2

conductor carrying current I_1 is the magnetic induction equal to zero?

5.2. Two mutually perpendicular conductors carrying currents I_1 and I_2 lie in one plane. Find the locus of points at which the magnetic induction is zero.

5.3. Equal currents are flowing along three conductors: a ring of radius R (Figure (a)), an infinitely long straight

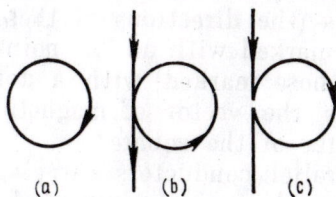

Fig. 5.3

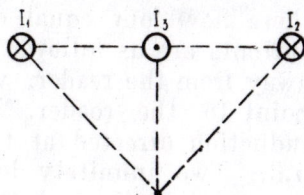

Fig. 5.4

conductor that forms a loop of the same radius R (Figure (b)), and an infinitely long straight conductor that also forms a loop of radius R but is broken at the point where the loop touches the conductor (Figure (c)). Find the relationships that link the magnetic induction vectors at the center of each circle.

5.4. Three conductors carrying currents are perpendicular to the plane of the drawing. They intersect the plane at three points that lie on a single straight line, with the distances from the middle conductor to the other two being equal. The currents in the outer conductors flow away from the reader, while the current in the middle conductor flows toward the reader. How is the magnetic field vector directed at the point on the straight line that is perpendicular to the straight line passing through the

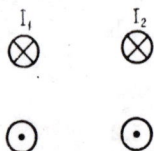

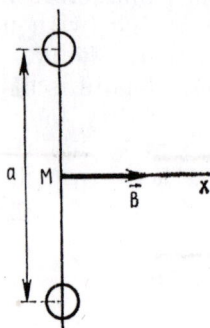

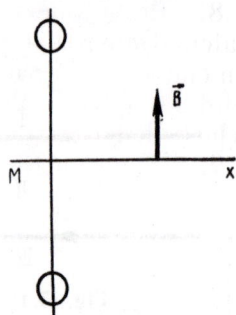

Fig. 5.5   **Fig. 5.6**   **Fig. 5.7**

three conductors in the plane of the drawing and is separated from the middle conductor by a distance equal to the distances between that conductor and the outer conductors? All three currents are equal in magnitude.

5.5. Along four parallel conductors whose sections with the plane of the drawing lie at the vertices of a square there flow four equal currents (the directions of these currents are as follows: those marked with an "x" point away from the reader, while those marked with a dot point to the reader. How is the vector of magnetic induction directed at the center of the square?

5.6. Two infinitely long parallel conductors carrying currents are directed at right angles to the plane of the drawing. The maximum of magnetic induction is at a point M that lies in the middle between the conductors. The direction of the magnetic induction vector **B** at this point coincides with the positive direction on the x axis. Determine the direction of the currents flowing in the conductors and the relationship that exists between these currents.

5.7. Two infinitely long parallel conductors carrying currents are directed at right angles to the plane of the drawing. The magnetic induction at a point M that lies in the middle between the conductors is zero. To the right of this point, the magnetic induction vector points upward, at right angles to the x axis. Find the direction of the currents flowing in the conductors, the direction of the magnetic induction vector to the left of point M, the relationship between the currents, and the point on the x axis at which the magnetic induction is maximal. The distance between the conductors is a.

5.8. Prove solely by reasoning (without performing any calculations) that the magnetic induction on the axis at an end face of a very long solenoid is half the value in the middle of the solenoid. A "very long solenoid" is one whose length is much greater than the diameter.

5.9. A current flows clockwise in a flat square loop. In

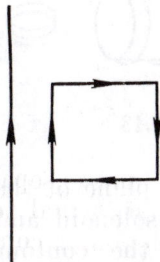 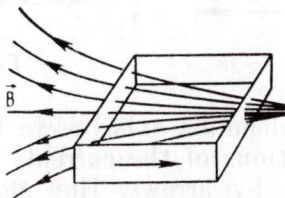 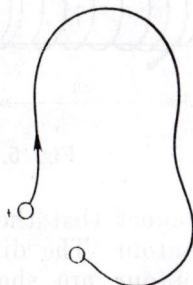

Fig. 5.9   Fig. 5.10   Fig. 5.11

the plane of the loop there lies an infinitely long straight conductor carrying a current whose direction is designated by the arrow in the figure. How will the loop move in the magnetic field created by the current flowing in the straight conductor and how will the shape of the loop change as a result of the action of this field?

5.10. A conducting loop carrying a current is placed in a nonuniform magnetic field. How will it move as a result of the action of this field?

5.11. A direct current (constant in magnitude and direction) flows in a contour made from soft wire. What shape does this contour tend to acquire as a result of the action of the magnetic field created by the current?

5.12. A small flat contour with a current flowing in it is placed successively at three points on the axis of a solenoid in which a current also flows in the same direction.

The points are at the middle of the solenoid (point *1*), at an end face (point *2*), and outside the solenoid at a distance from an end face equal to one-half the length of the solenoid (point *3*). The plane of the contour and the plane of the cross section of the solenoid are parallel. At which of these three points does the contour experience the greatest interaction with the solenoid and at which is the force minimal? Is the force attractive or repulsive at these points? The length of solenoid is considerably larger than the diameter.

5.13. At a small distance from a solenoid carrying a current there is placed a contour with a current in such a

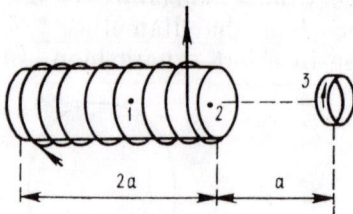
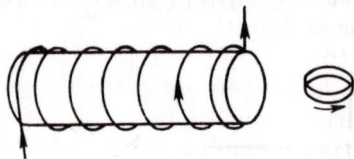

Fig. 5.12   Fig. 5.13

manner that the solenoid's axis lies in the plane of the contour. The directions of the currents in solenoid and contour are shown by arrows. How does the contour move? How will it move if the current in it flows in the direction opposite to the one shown in the figure?

5.14. Between two fixed contours, *1* and *3*, carrying currents that flow in the same direction there is suspended

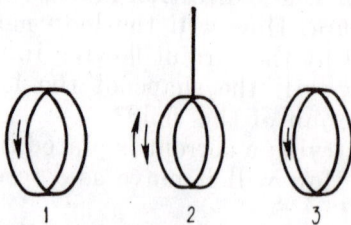
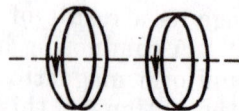

Fig. 5.14   Fig. 5.15

another contour, *2*, that also carries a current. Contour *2* is oriented in such a manner that the forces caused by the currents in contours *1* and *3* are opposite in direction, equal in magnitude, and lie along a single straight line; thus, contour *2* is in equilibrium. Is this state of equilib-

70

rium stable or unstable? Consider the case where the current in contour *2* has the same direction as the currents in *1* and *3* and the case where the directions are opposite.

5.15. Two contours whose planes are parallel to each other and are separated by a certain distance carry currents that flow in the same direction. One contour is left fixed while the other is positioned in a different manner with respect to the first: in one case its plane is turned by 90°, in the other by 180°, while in the third case it is just moved parallel to itself over a certain distance. In which of these three cases one will have to perform the greatest work and in which, the smallest?

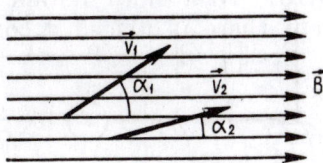

Fig. 5.16

5.16. In a uniform magnetic field there are two charged particles moving with velocities v_1 and v_2 and carrying equal charges, with $|\,v_1\,| = |\,v_2\,| = v$. The velocity of one particle forms an angle α_1 with the direction of the field, while the other velocity forms an angle α_2. In what parameters does the motion of one particle differ from that of the other? Determine which of the parameters is greater for which particle.

5.17. The device shown in Figure (a) is commonly used to measure the charge-to-mass ratio of the electron. The electrons that leave the cathode *C* are accelerated by an electric field that exists in the space between the cathode and the anode *A*. A fraction of electrons fly through the hole in the anode. These electrons, leaving region *I* of the device, fly into the region where there is no electric field. In this region the electrons are deflected from a straight line via a magnetic field directed perpendicularly to the plane of the drawing. This field is generated by two solenoids. The region *II* where the trajectory of the electrons is bent lies between these two solenoids. By increasing the current flowing through the two solenoids connected in series we can direct the electrons into a Faraday cylinder *F*, with a galvanometer *G* registering the resulting current. Any further increase in the solenoid current results in a drop in the current flowing through *G*, since the electrons begin to move along a circle of a smaller radius. The dependence of the galvanometer current on the solenoid current is illustrated by the curve in Figure (b) The following quantities are known in measurements: the potential difference U between anode and cathode, the curvature radius R of the axial line of region II (assuming that the majority of electrons deflected by the magnetic field travel along this line), the number of turns N_0 per unit length of solenoid, and the solenoid current I at which the galvanometer current is maximal. How to determine the charge-to-mass ratio of the electron knowing the values of these quantities?

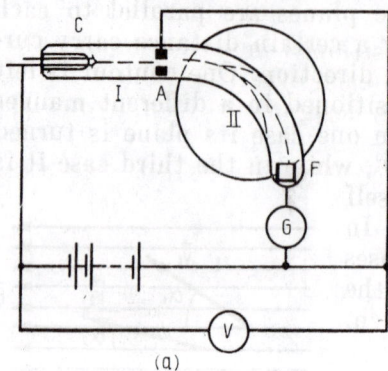

(a)

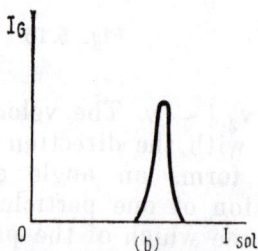

(b)

Fig. 5.17

5.18. A charged particle of mass m and charge Q has acquired a certain velocity by passing through a potential difference U_0. With this velocity it flies into the field of a parallel-plate capacitor, with the distance between the plates being l, the potential difference being U. The veloc-

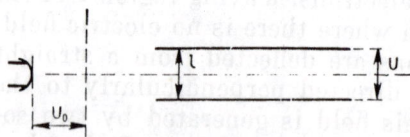

Fig. 5.18

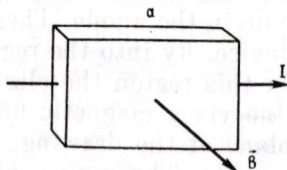

Fig. 5.19

ity of the particle is directed parallel to the plates. Where should the magnetic field that makes the particle move along a straight line in the capacitor be directed and what should its value be (the induction B)?

5.19. A direct current I is flowing through a plane in the direction designated by an arrow. The plate is placed in a transverse magnetic field **B**. As a result of the Hall effect there appears a transverse potential difference. What is the sign of the potential at point a if the plate is made of metal and if the plate is an n-type or p-type semiconductor?

5.20. Two contours are positioned in such a manner that their planes are parallel to each other. Contour *1* carries a current whose direction is designated by an arrow. The

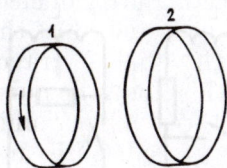

Fig. 5.20

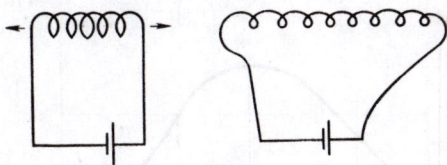

Fig. 5.21

contours move in relation to one another, but their planes remain parallel in the process. What is the direction of the current induced in contour *2* when the contours are moved toward each other or away from each other?

5.21. A spiral made from elastic wire is connected to a DC source. The spiral is stretched. Will the current flowing in the spiral become greater or smaller in the stretching process than the initial current or will it remain unchanged?

5.22. A solenoid carrying a current supplied by a DC source with a constant emf contains an iron core inside

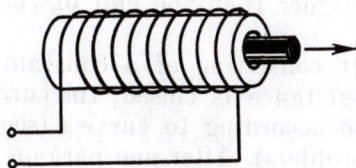

Fig. 5.22

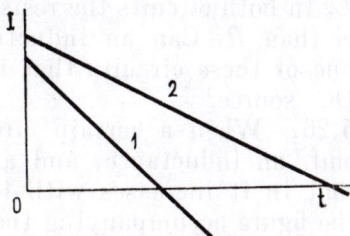

Fig. 5.23

it. How will the current change when the core is pulled out of the solenoid: will it increase, decrease, or remain the same?

5.23. Two identical inductances carry currents that vary with time according to linear laws. In which of the two inductances is the self-induction emf greater? Will the values or signs of the self-induction emf's change if the currents begin to increase in the opposite direction after they pass through zero (with the linear laws retained in the process)?

5.24. A current that varies with time according to a law depicted graphically in the figure passes through an induction coil. In which of the moments denoted in the figure

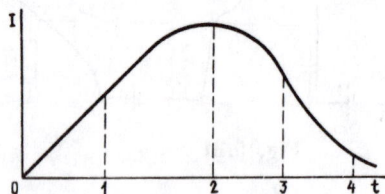 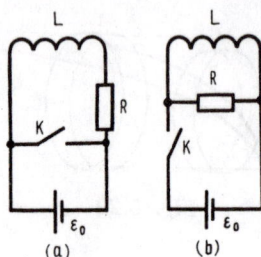

Fig. 5.24   Fig. 5.25

will the self-induction emf be maximal (the inductance of the coil remains unchanged in the process)?

5.25. Various circuits are used to observe the phenomenon of self-induction. Among these are the circuits shown in Figures (a) and (b). In Figure (a), key K is initially opened and the current flows through the induction coil L and resistor R connected in series. In Figure (b), key K is initially closed and the current branches off to R and L. In both circuits the resistance of the coil L is much lower than R. Can an induction emf be generated in either one of these circuits that is higher than the emf of the DC source?

5.26. When a certain circuit consisting of a constant emf, an inductance, and a resistance is closed, the current in it increases with time according to curve 1 (see the figure accompanying the problem). After one parameter ($\mathscr{E}$, L, or R) is changed, the increase in current follows curve 2 when the circuit is closed a second time. Which parameter was changed and in what direction?

5.27. A current is flowing in a circular contour 1 whose radius is R. A second contour, 2, whose radius is much smaller than that of the first, is moving with a constant velocity $\mathbf{v}$ along the r axis in such a manner that the planes of the contours remain parallel to each other in the course of the motion. At what distance from contour *1* will the emf induced in contour *2* be maximal?

5.28. A certain circuit consists of a DC source with emf $\mathcal{E}$, an induction coil *L1*, and a key *K1*. No resistance is present in the circuit. Another coil, *L2*, which is connected electrically to a resistor *R* through a key *K2*, is fastened to *L1*. At some moment in time key *K1* is closed. After a certain time interval *K2* is closed. How do

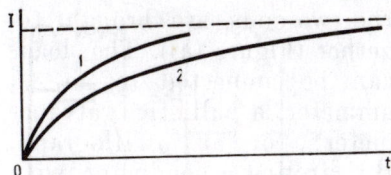

Fig. 5.26

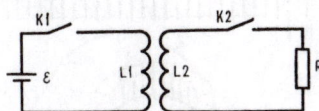

Fig. 5.28

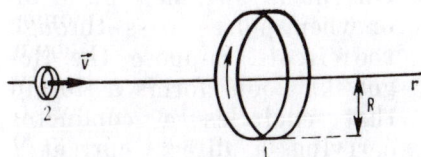

Fig. 5.27

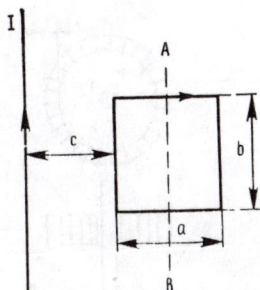

Fig. 5.29

the current in the primary circuit (the one containing $\mathcal{E}$), the induction emf in the secondary circuit (the one with *L2* and *R*), and the current in the secondary circuit vary with time?

5.29. An infinitely long straight conductor and a flat rectangular contour with sides *a* and *b* and with *N* turns lie in a single plane. The distance between the straight conductor and the side of the contour closest to the straight conductor is *c*. Determine the following quantities: (1) the mutual inductance of the conductor and the contour; (2) the quantity of electricity induced in the contour if the contour is rotated through 90° about the *AB* axis provided that a current *I* is flowing in the contour and the resistance of the contour is *R*; (3) the work that must be done to rotate the contour through 180° about the AB axis provided that there is current I both in the long conductor and in the contour and that the sense of the current in the contour is clockwise (in the plane of the drawing).

5.30. A common device used in electrical measurements is the so-called Rogowski loop. It constitutes a flexible solenoid that can be transformed into a torus if the two ends are brought together (Figure (a)). The leads can be connected to an AC ammeter, a ballistic galvanometer*, or an oscillograph. By circling a conductor with a Rogowski loop one can measure an alternating current flowing constantly in the conductor or even isolated changes in the current, such as those that occur when the current is switched on or off or when pulses pass through the circuit. Suppose the Rogowski loop forms a toroid that encircles a conductor carrying a direct current I (Figure (b)). The parameters of the loop are as follows: the cross-sectional area is S, the number of turns is N,

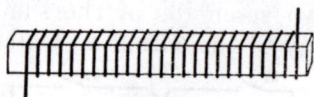

(a)

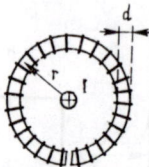

(b)

Fig. 5.30

the resistance of the winding is R, and the radius of the toroid is r. It is assumed that the width d of the loop proper is very small compared to r. At a certain moment the current is switched off; the current becomes zero in a very short interval. The ballistic galvanometer in the circuit of the loop measures the quantity of electricity Q that has passed through the loop (and the galvanometer). How can one find the current I that was flowing in the conductor prior to switch-off knowing the values of the above-mentioned parameters?

* A ballistic galvanometer has a large period of oscillations. It is commonly used to measure the quantity of electricity that flows in a circuit in the form of a short pulse.

5.31. A flat coil with a cross-sectional area S and with N turns is placed in a magnetic field. The leads of the coil are connected to an oscillograph. When the coil is moved out of the field, an induction emf is generated in it, and the oscillogram of this emf is shown in the figure. How do the maximal value of the emf, $\mathscr{E}_{lm}$, and the area under the curve depend on the rate with which the coil is moved out of the field?

5.32. Suppose that we have two solenoids of the same length. Their diameters differ only to the extent to which

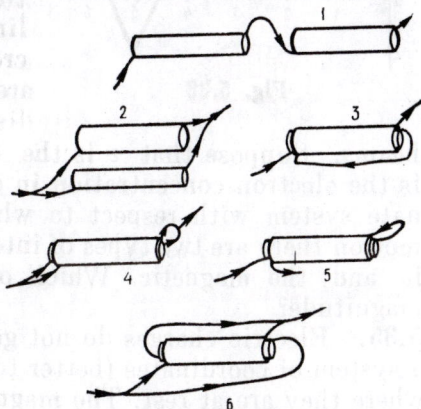

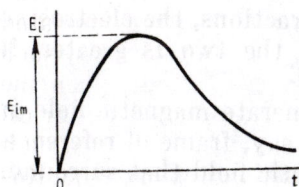

Fig. 5.31   Fig. 5.32

one can be fitted onto the other. The inductances of the two solenoids can be considered the same and equal to L. Here are the ways in which the solenoids can be connected:

(1) the solenoids are connected in series and are separated by a large distance;

(2) the solenoids are connected in parallel and are separated by a large distance;

(3) the solenoids are connected in series, one is fitted onto the other, and the senses of the turns coincide;

(4) the solenoids are connected in parallel, one is fitted onto the other, and the senses of the turns coincide;

(5) the solenoids are connected in series, one is fitted onto the other, and the senses of the turns are opposite;

(6) the solenoids are connected in parallel, one is fitted onto the other, and the senses of the turns are opposite.

Determine the total inductance for each of the above cases.

5.33. The current flowing in a certain inductance coil varies in time according to the curve shown schematically in the figure. Draw the curve representing the induced emf as a function of time (also schematically).

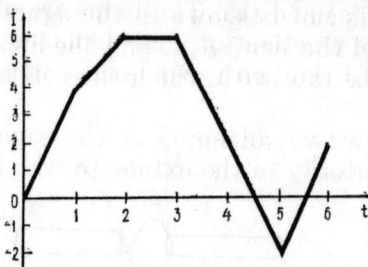

Fig. 5.33

5.34. Two similar parallel electron beams point in the same direction. The linear dimensions of the cross section of each beam are small compared to the distance between the beams. Suppose that v is the electron velocity and n is the electron concentration in either beam. In a coordinate system with respect to which the electrons are in motion there are two types of interactions, the electrostatic and the magnetic. Which of the two is greater in magnitude?

5.35. Electric charges do not generate magnetic field in a system of coordinates (better to say, frame of reference) where they are at rest. The magnetic field that surrounds a conductor carrying a current is generated by the charges that are moving in the conductor. Since the electron concentration in a conductor is of the order of 10^{22} cm^{-3}, the directional velocity of the electrons in the conductor is of the order of one millimeter per second (if the current density is estimated at 100 A/cm^2). We position the conductor carrying the current in such a manner that it follows the magnetic meridian at the point where the conductor is present. Just as in Oersted's experiment, a magnetic compass needle placed under the conductor will be deflected. If the needle is moved along the conductor with a speed equal to the directional velocity of the electrons in the conductor (i.e. of the order of several millimeters per second), the electron will be at rest in relation to the needle and, since the magnetic field in the system connected with the needle must be nil, the needle will not be deflected. More than that, if the needle is moved along the conductor with a speed greater than that of the electrons, the needle will be deflected in the opposite direction. Are these assertions correct?

5.36. How are the magnetic induction vector and the magnetic field vector directed inside and outside a bar magnet?

5.37. Two types of steel are characterized by the hysteresis loops shown in Figure (a) and (b). The loops are obtained in the processes of magnetization and demagnetization of the steels. Which of the two types is best suited

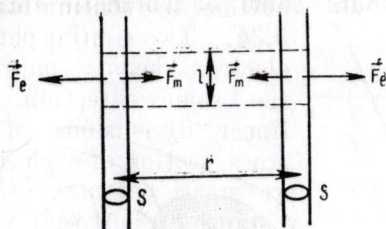

Fig. 5.34

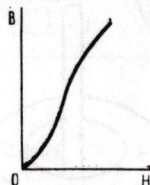

Fig. 5.38

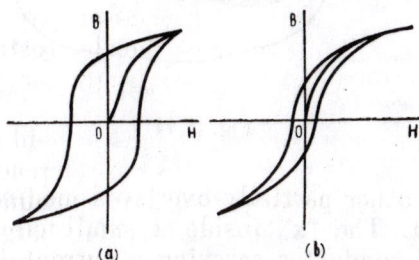

(a)   (b)

Fig. 5.37

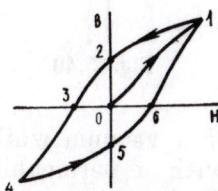

Fig. 5.39

for using as the core of a transformer and which, for using as a permanent magnet?

5.38. How can one use the B vs. H graph (the magnetization curve) to determine the work that a source of current must perform to magnetize a ferromagnetic core of a solenoid whose length is l and whose cross-sectional area is S? The magnetization curve is shown in the figure accompanying the problem.

5.39. Does a hysteresis loop possess sections in which we can formally assign to permeability a value that is zero or infinite or negative?

5.40. A straight conductor passes through a ferromagnetic toroid, as shown in the figure accompanying the problem. The conductor carries a current that first grows to a certain maximal value and then falls off to zero, as a result of this the toroid becomes magnetized. Indicate the directions of the lines of force for magnetic induction in the toroid and find the sections or points on the hysteresis loop corresponding to the state of the toroid after the current has ceased to flow (see the figure accompanying Problem 5.39).

5.41. Suppose we wish to calculate the circulation integrals of the magnetic field strength and magnetic induction along various contours, some of which lie entirely

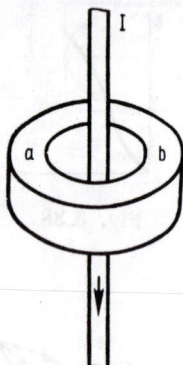

Fig. 5.40

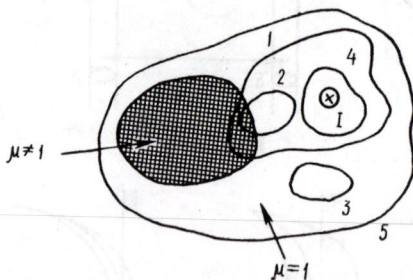

Fig. 5.41

in a vacuum while the other partially overlap a medium with a permeability μ. The "x" inside a small circle marks the section of a conductor carrying a current by the plane of the drawing. Are all the circulation integrals of the magnetic induction equal to each other? Is this also true of the circulation integrals of the magnetic field strength?

6. Oscillatory Motion and Waves

6.1. At two moments in time the displacements of a harmonically oscillating point are the same. Can we state, on the basis of what we have just said, that the phases at these moments are also the same?

6.2. The oscillations depicted by curve *1* in the figure are expressed by the equation $x = A \sin \omega t$. What is the equation for the oscillations depicted by curve *2*?

6.3. Two material particles of equal mass are performing harmonic oscillations whose graphs are shown in the figure. What oscillation has a higher energy?

6.4. As a result of adding two mutually perpendicular oscillations of equal frequency, the motion of an object occurs along an ellipse; in one case the motion is clockwise, while in the other it is counterclockwise. Write the

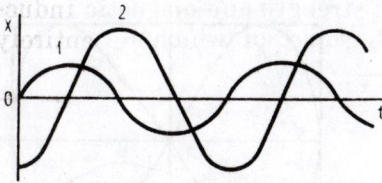

Fig. 6.2

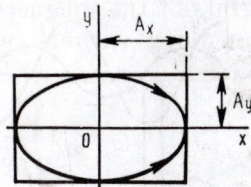

Fig. 6.4

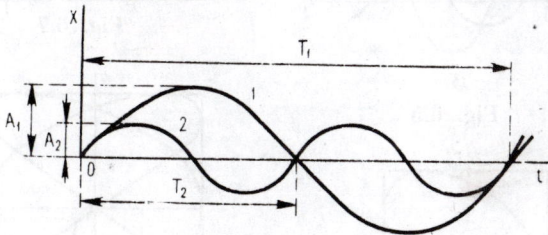

Fig. 6.3

equations of motion along each coordinate axis, assuming that the initial phase along the x axis is zero.

6.5. Two mutually perpendicular oscillations are added. In one case the graphs representing these oscillations are those shown in Figure (a) and in the other, those shown in Figure (b). In what respect do the resultant oscillations differ?

6.6. Suppose that the addition of two mutually perpendicular oscillations in which a material particle participates results in an ellipse, with the direction of motion indicated by the arrow in the figure. The equation of motion along the x axis can be written in the form $x = A_1 \sin \omega t$ and that along the y axis, in the form $y = A_2 \sin (\omega t + \varphi)$. Determine the condition that φ must meet.

6.7. Two mutually perpendicular oscillations obey the laws

$$x = A_1 \sin \omega_1 t \text{ and } y = A_2 \sin (\omega_2 t + \varphi).$$

The addition of these two oscillations leads to the Lissajous figure shown in the drawing accompanying the problem. Determine the relationship between ω_1 and ω_2 and the initial phase φ if the figure is traversed in the direction shown by the arrows.

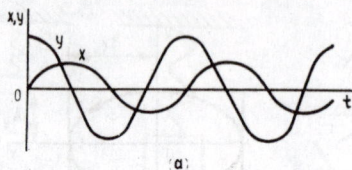

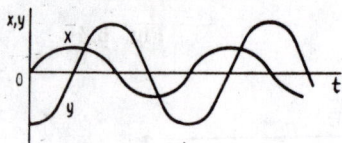

Fig. 6.5

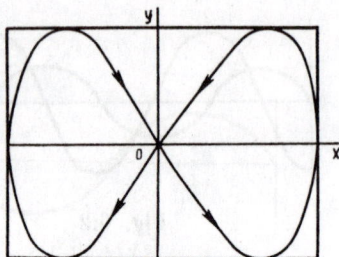

Fig. 6.7

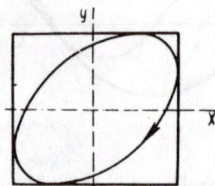

Fig. 6.6

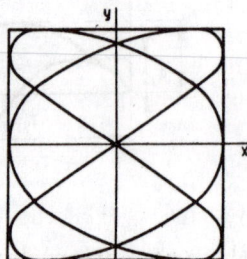

Fig. 6.8

6.8. Two mutually perpendicular oscillations are performed according to the laws

$$x = A_1 \sin \omega_1 t \quad \text{and} \quad y = A_2 \sin \omega_2 t.$$

Determine the relationship between ω_1 and ω_2 using the Lissajous figure shown in the drawing accompanying the problem.

6.9. A material particle oscillates according to the harmonic law. At which of the two moments, *1* or *2*, is the kinetic energy higher and in which, the potential energy? At which moment is the acceleration of the particle at its maximum (in absolute value)?

6.10. Two loads whose masses are m_1 and m_2 are suspended by springs ($m_1 > m_2$). When the loads were attached to the unloaded springs, it was found that the elongations of the springs were the same. Which of the two loads oscil-

lates with a greater oscillation period and which of the
two loads possesses a higher energy (provided that the
oscillation amplitudes are equal)? The springs are considered massless.

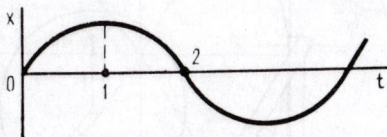

Fig. 6.9

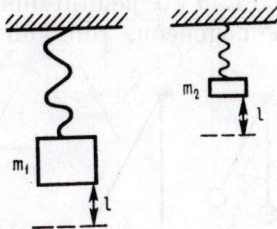

Fig. 6.10

6.11. A chemical test tube is balanced by a load at its
bottom so that it does not tip when submurged in a liquid
(the cross-sectional area of the tube is S). After submerg-
ing to a certain depth, the tube begins to oscillate about
its position of equilibrium. The tube, whose mass together with the mass of the load is m, is in the state of equilibrium in a liquid with a density ρ when its bottom is below the level of liquid by a distance l. Determine the oscillation period of the tube assuming that the viscosity of
the liquid is nil.

6.12. One way to measure the mass of an object in a space
station at zero gravity is to use a device schematically

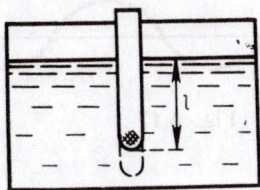

Fig. 6.11

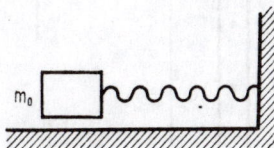

Fig. 6.12

shown in the figure. The principle of operation of this
device is as follows. First the austronaut measures the
oscillation frequency of an elastic system of known mass.
Then the unknown mass is added to this system and a new
measurement of the oscillation frequency is taken. How
can one determine the unknown mass from the two measured values of frequency?

6.13. Two simple pendulums having equal masses but different lengths are in oscillatory motion with the same angular amplitudes. Which of the two pendulums has a higher oscillation energy?

6.14. Two pendulums, a physical one in the form of a homogeneous rod and a simple one, of equal mass and

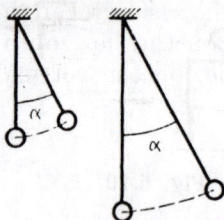

Fig. 6.13

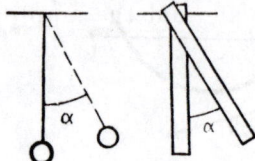

Fig. 6.14

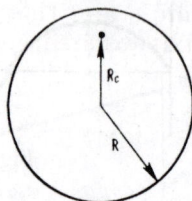

Fig. 6.15

length are in oscillatory motion with the same angular amplitudes. Which of the two pendulums has a higher oscillation energy?

6.15. An axis passes through a disk of radius R and mass m at a distance R_c from the disk's center. What will be the period of oscillations of the disk about this axis (which is fixed)?

6.16. Consider a physical pendulum that is a homogeneous rod of length l. At what distance R_c from the center of

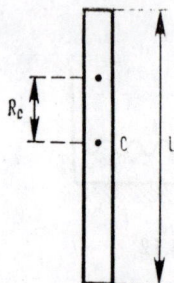

Fig. 6.16

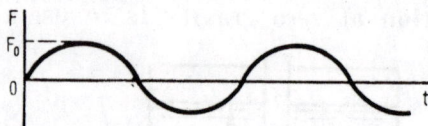

Fig. 6.17

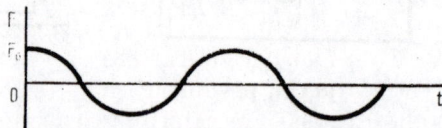

Fig. 6.18

gravity of the rod must the point of suspension lie for the oscillation period to be maximal?

6.17. A force acting on a material particle varies according to the harmonic law

$$F = F_0 \sin \omega t.$$

At time $t = 0$ the velocity v is zero. How do the velocity and position of the particle vary with time?

6.18. A force acting on a material particle varies according to the harmonic law

$$F = F_0 \cos \omega t.$$

At time $t = 0$ the velocity v is zero. How do the velocity and position of the particle vary with time?

6.19. The time dependence of the amplitude of damped oscillations is presented in the figures on a semilogarithmic scale, that is, the time is laid off on the horizontal axis on a linear scale and the amplitude, on the vertical

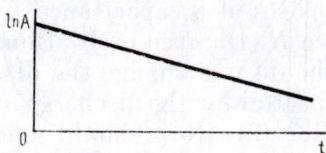

Fig. 6.19

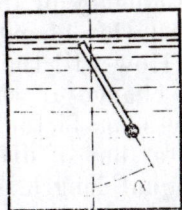

Fig. 6.21

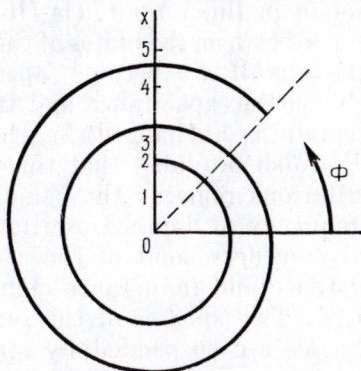

Fig. 6.20

axis on a logarithmic scale. Construct the time dependence of the energy of these oscillations using the semilogarithmic scale. Set the initial values of the logarithms of the amplitude and energy of the oscillations equal.

6.20. Suppose that certain damped oscillations are represented in polar coordinates. Depict these oscillations in Cartesian coordinates with the phase of the oscillations laid off on the horizontal axis and the displacement, on the vertical axis, assuming that the ratio of the sequential amplitudes of oscillations and the initial phase remain unchanged. Find the logarithmic decrement of the oscillations.

6.21. Suppose that a pendulum oscillates in a viscous medium. The viscosity of the medium and the mass and length of the pendulum are such that the oscillations are aperiodic. The pendulum is deflected from the position of equilibrium and released. How will the absolute value of the the pendulum's velocity vary with time: will it increase continuously, decrease continuously, pass through a maximum, or pass through a minimum?

6.22. A load suspended by a spring in a viscous medium performs damped oscillations. How should one change the length of the spring (preserving all the characteristics of the spring, i.e. the thickness of the wire, the density of the turns, etc.) so that the oscillations become aperiodic? The mass of the spring is assumed to be negligible compared to the mass of the load.

6.23. An oscillatory circuit consists of a capacitance C, an inductance L, and a resistance R. Damped oscillations set in in this circuit. (1) How should one change the distance between the plates of the capacitor for the discharge in the circuit to become aperiodic? (2) How should one change the capacitance and inductance (with the resistance remaining unchanged) for the damping in the contour to diminish provided that the natural frequency of free oscillations remains the same? How will this change the frequency of damped oscillations? (3) How will the logarithmic decrement of the oscillations change if the resistance and inductance change by the same factor?

6.24. Two spheres of the same diameter but of different masses are suspended by strings of equal length. If the

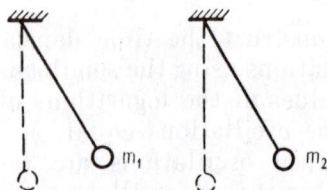

Fig. 6.24

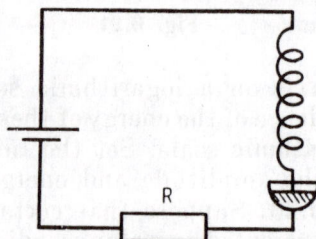

Fig. 6.25

spheres are deflected from their positions of equilibrium, which of the two will have a greater oscillation period and which will have a greater logarithmic decrement if their oscillations occur in a real medium with viscosity?

6.25. A "dancing spiral" is sometimes demonstrated at lectures. A spring fixed at its upper end is submerged by its lower end into mercury. Voltage supplied by a DC

source is applied to the upper end and the mercury.
When current flows in the spring, the rings of the spring
tend to draw together, the spring gets shorter, and the
lower end moves out of the mercury. The current ceases,
and the lower end is again submerged in the mercury. The
process repeats itself. What oscillations does the spring
perform in the process: free, forced, damped, or self-oscillations?

6.26. Which of the two diagrams, Figure (a) or Figure
(b), represents the dependence of the amplitude of displacements in forced oscillations on the frequency of the driving force and which represents the frequency dependence

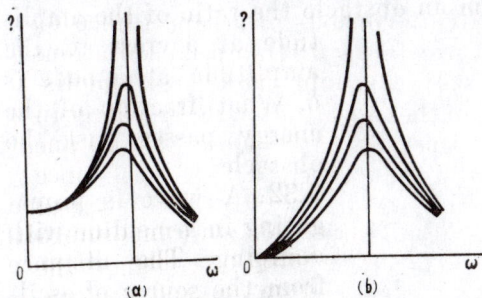

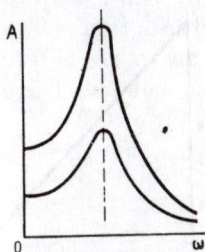

Fig. 6.26   Fig. 6.29

of the velocity amplitude? In what parameter determining
the oscillation conditions does each curve represented in
Figures (a) and (b) differ? What parameters determine the
intersection of each curve with the vertical axis in Figure
(a) and the position of the maximum?

6.27. How will the displacement amplitude at $\omega = 0$
that is A_0, the maximal amplitude A_m, and the resonance
frequency ω_{res} vary if the resistance of the medium in
which the oscillations occur decreases provided that all
the other parameters that determine the forced oscillations remain unchanged?

6.28. The curve depicting the dependence of the amplitude of forced oscillations on the frequency of the driving
force in a medium with no resistance tends to infinity as
$\omega = \omega_0$. Why is this situation meaningless not only from
the physical standpoint but also from the mathematical
standpoint? How does a system oscillate in a medium
that has practically no resistance?

6.29. Two forced oscillations with the same natural frequencies have amplitudes that differ by a factor of 2 for

all values of the frequency of the driving force. In what parameter, among the amplitude of the driving force, the mass of the oscillating object, the elasticity coefficient, and the resistance of the medium, do these systems differ? It is assumed that these systems may differ only in one parameter.

6.30. Waves on the surface of water in the form of parallel lines advance on a wall with an aperture much narrower than the wavelength. What will be the shape of the waves propagating on the surface behind the wall (and aperture)?

6.31. In the standing waves that form as a result of reflection of waves from an obstacle the ratio of the amplitude at a crest to the amplitude at a node is δ. What fraction of the energy passes past the obstacle?

6.32. A wave is propagating in a medium with damping. The distance from the source of oscillations (in units of the wavelength) is laid off on the horizontal axis and the common logarithm of the oscillation amplitude is laid off on the vertical axis. Using the graph shown in the figure accompanying the problem, write a formula that will link the amplitude with the distance.

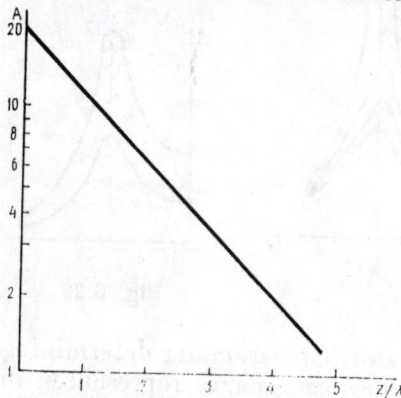

Fig. 6.32

6.33. The formula that expresses the speed of sound in a gas can be written in the following form:

$$c = \sqrt{\gamma p / \rho}. \qquad (6.33.1)$$

Here γ is the specific heat ratio (the ratio of the specific heat capacity of the gas at constant pressure to the specific heat capacity at constant volume), p is the pressure of the gas, and ρ is the density of the gas. Using this formula as a basis, can we stipulate that upon isothermal change of the state the speed of sound in the gas grows with pressure?

88

6.34. The figure demonstrates the temperature dependence of the speed of sound in neon and water vapor on the log-log scale. Which straight line corresponds to the lighter of the gases?

6.35. The dependence of the frequency of oscillations registered by a receiver when the receiver and the source of sound approach each other depends on whether the source moves and the receiver is fixed, or whether the source is fixed and the receiver is in motion. The curves in the figure represent the dependence of the ratio of the received

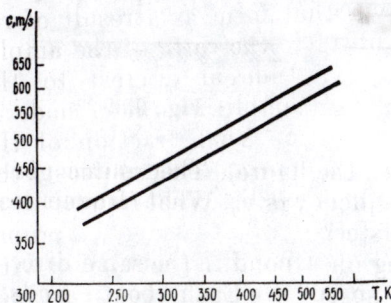 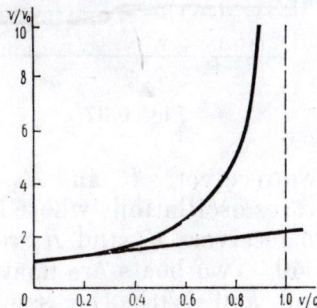

Fig. 6.34   Fig. 6.35

frequency of oscillations to the frequency emitted by the source on the ratio of the rate of relative motion to the velocity of sound. Which of the two curves corresponds to a moving source and which, to a moving receiver? The medium where the propagation of sound takes place (air or water) is assumed fixed.

6.36. An observer standing at the bed of a railroad hears the whistle of the locomotive of the train that rushes past him. When the train is approaching the observer, the frequency of the whistle sound is ν_1, while when it has passed the observer, the frequency is ν_2. Determine the speed of the train and find the whistle frequency when the observer moves together with the train. The speed of sound is assumed to be known.

6.37. Two observers stand at different distances from the bed of a railroad. When a train passes them, each hears how the frequency of the train whistle changes, with the change occurring along curve 1 for one observer and along curve 2 for the other. Which of the two observers is standing closer to the roadbed?

89

6.38. A source of sound whose frequency is ν_0 is moving with a speed v. The waves travel to a fixed obstacle, are reflected by the obstacle, and are registed by a receiver that moves together with the source. What frequency is registed by the receiver if the speed of sound waves is c?

6.39. A source of oscillations S is fixed to the riverbed of a river whose waters flow with a velocity v. Up and down the stream there are fixed (also to the river bed)

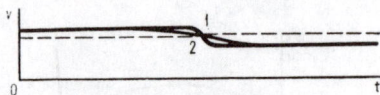

Fig. 6.37

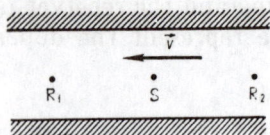

Fig. 6.39

two receivers, R_1 and R_2 (see the figure). The source generates oscillations whose frequency is ν_0. What frequencies do receivers R_1 and R_2 register?

6.40. Two boats are floating on a pond in the same direction and with the same speed **v**. Each boat sends, through the water, a signal to the other. The frequencies ν_0 of the generated signals are the same. Will the times

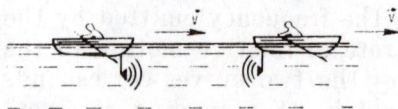

Fig. 6.40

Fig. 6.41

it takes the signals to travel from one boat to the other be the same? Are the wavelengths the same? Are the frequencies received by the boats the same?

6.41. An underground explosion at a point A generates vibrations. Seismographs that are capable of measuring longitudinal and transverse waves separately are placed at another point B. The time interval between the arrival of longitudinal and transverse waves is measured. How, knowing the velocities of propagation of longitudinal and transverse waves and the time difference between arrival, to determine the distance S between points A and B?

6.42. A sound wave travels in air and falls on the interface between air and water at an angle α_1. At what angle

α_2 will the wave propagate in the medium: greater than α_1 or smaller than α_1?

6.43. There are many documented cases when an explosion at a point A will be heard at a point B that is located far away from A while in a certain region, known as the zone of silence, located much closer to A than to B the explosion is not heard. Among the reasons for this is the deflection of sound waves caused by the presence of a vertical temperature gradient in the atmopshere. How

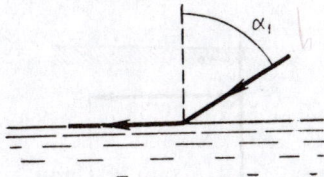

Fig. 6.42

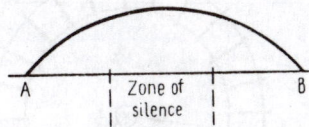

Fig. 6.43

should the air temperature change with altitude for the direction of propagation of sound waves to be as shown in the figure?

6.44. At a depth h_1 below ground level there is a pocket of water of depth h_2. What type of artificial seismic waves, longitudinal or transverse, is needed to measure the depth of the water pocket?

6.45. An airlane is in supersonic flight at an altitude h. At what smallest distance a (along the horizontal) from

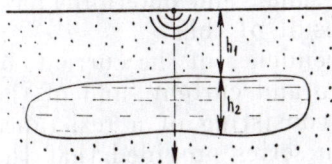

Fig. 6.44

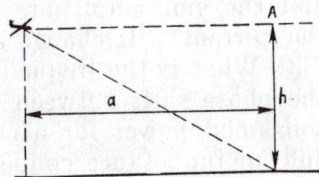

Fig. 6.45

the observer on the ground is there a point from which the sound emitted by the airplane motors travels to the observer faster than from point A that is directly above the observer?

7. Alternating Current

7.1. Using a Rogowski loop (see Problem 5.30), one can measure the effective value I_{eff} of an alternating current flowing in a conductor. The loop has a rectangular cross section with N turns. The dimensions and the position of the loop are shown in the figure. Determine the effective emf generated in the loop by the alternating current.

7.2. The figure shows the vector diagram of reactances and resistances in an AC circuit. Construct a similar dia-

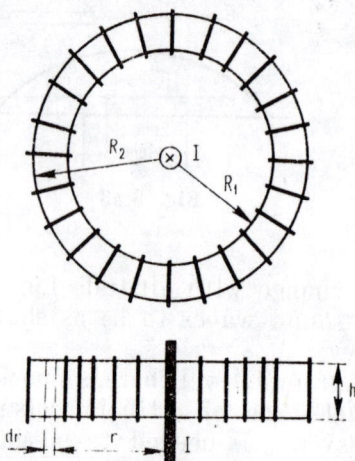

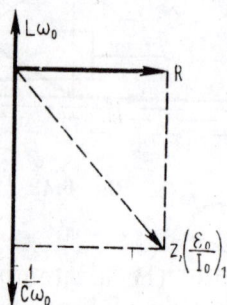

Fig. 7.1   Fig. 7.2

gram for a circuit in which the current frequency is doubled and the emf amplitude is the same, and determine how the current will change as a result of this.

7.3. What is the frequency dependence of the current, of the phase shift between voltage and current, and of the consumed power for a circuit consisting of a resistance and an inductance connected in series provided that the emf amplitude remains constant?

7.4. What is the frequency dependence of the current, of the phase shift between current and voltage, and of the consumed power for a circuit consisting of a resistance and a capacitance connected in series provided that the emf amplitude remains constant?

7.5. A circuit (Figure (a)) contains an alternating emf, a resistance, and a **reactive** element (only a capacitance or only an inductance). What is this element if the time dependences of the current in the circuit and the emf of the source are those as shown in Figure (b)?

7.6. Are the readings of the ammeter *A3* equal to the sum of the readings of the ammeters *A1* and *A2* for the cases depicted in Figures (a) and (b)?

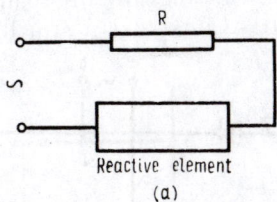

Reactive element

(a)

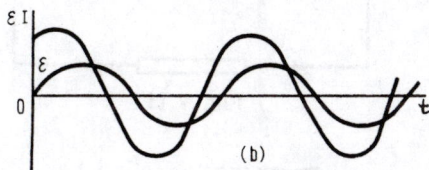

(b)

Fig. 7.5

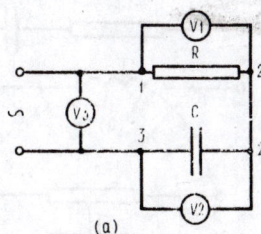

(a)

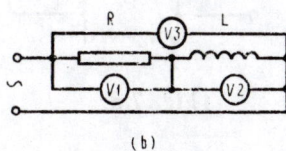

(b)

Fig. 7.7

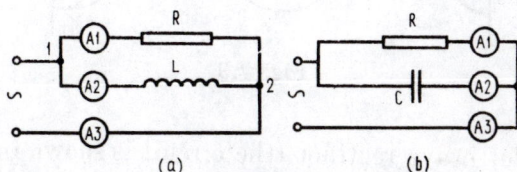

(a)   (b)

Fig. 7.6

7.7. Are the readings of the voltmeter *V3* equal to the sum of the readings of the voltmeters *V1* and *V2* for the cases depicted in Figures (a) and (b)?

7.8. The current flowing through the resistance in an AC circuit shown in Figure (a), where a resistance R, a capacitance C, and an inductance L are connected in series, is $I = \mathscr{E}/R$. What will be the current in the AC circuit when the inductance and the capacitance connected in parallel are connected in series with the resistance (Figure (b))?

7.9. The power in an AC circuit varies with time according to the curve in the figure. How, knowing the maximal and minimal values of the power, to determine the numerical value of the phase shift between voltage and

current? What is the period of variation of the power?
7.10. To demagnetize watches that have been accidentally magnetized, they are placed inside a solenoid connected to an AC source. The watches are then slowly removed from the solenoid. Explain why the watches become demagnetized as a result of such manipulations.

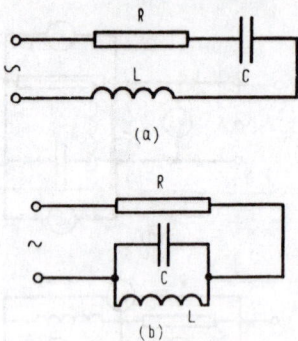

Fig. 7.8

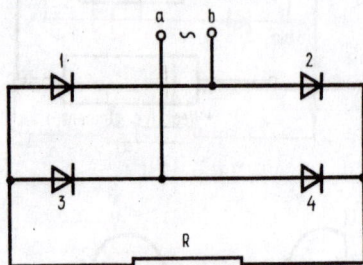

Fig. 7.11

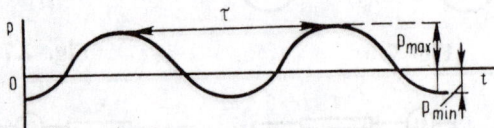

Fig. 7.9

7.11. A full-wave rectifier (the circuit is shown in the figure) rectifies the current that flows continuously in one direction. Sketch the time dependence of the current, ignoring all losses, and, assuming that the load of the rectifier constitutes a resistance, calculate the average value of the current. If the rectifier is loaded to a primary winding of a transformer, is a constant emf generated in the secondary winding?

7.12. In the circuit shown in the Figure, a capacitor of capacitance C is connected in parallel with a resistor R. How will this influence the time dependence of the current?

7.13. Two semiconductor diodes in opposition to each other in series are connected to the primary winding of a transformer. Draw the oscillograms of the current in the primary winding and of the emf generated in the secondary winding.

7.14. Two vacuum diodes in opposition to each other in parallel are connected to the primary winding of a transformer. The amplitude of the emf applied to the primary

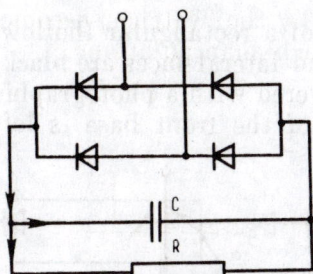

Fig. 7.12

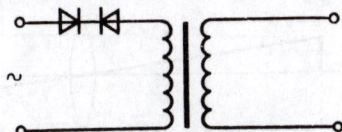

Fig. 7.13

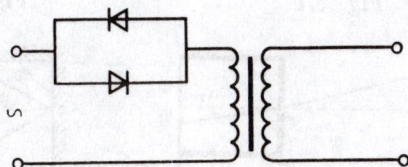

Fig. 7.14

winding exceeds considerably the voltage at which the diodes go into the saturation mode. Draw the oscillograms of the current in the primary winding and of the emf generated in the secondary winding.

8. Optics

8.1. At what distance f_1 from a biconvex lens must we place an object for the distance between the object and the real image to be minimal?

8.2. Two biconvex lenses a and b with the same radii of curvature are manufactured from glass samples with different refractive indices. How should we employ the graphs that represent the dependence of the distance f_2 between a lens and the image of an object on the distance f_1 between the lens and the object in order to determine the ratio of the refractive indices?

8.3. When taking a picture of a group of objects that are positioned at different distances from the camera, one must allow for the so-called depth of focus, or the limits

of the greatest and the smallest distance between which
the image is sharp for a given focus setting of the camera.
Why is the depth of focus the greater the smaller the
aperture setting?

8.4. A pinhole camera consists of a rectangular (hollow)
glass prism whose front base and lateral faces are blackened and whose back base is covered with a photographic
plate. A small circular section of the front base is left

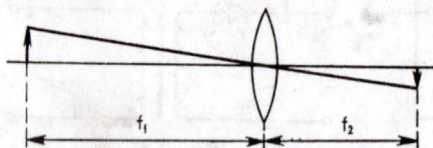

Fig. 8.1

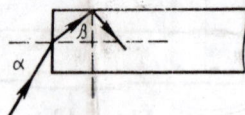

Fig. 8.5

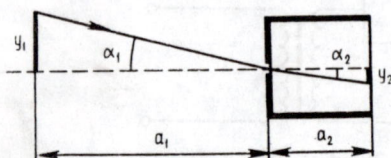

Fig. 8.4

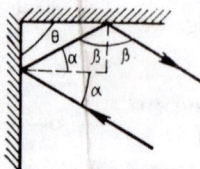

Fig. 8.6

unblackened, and through this "pinhole" the light enters
the camera. The refractive index of the glass is n, the distance from the object to the camera is a_1, and the length
of the camera is a_2. Determine the ratio of the size of the
image, y_2, to the size of the object, y_1, assuming that $y_1 \ll a_1$.

8.5. Light falls on an end face of a glass rod at an angle
α. What is the smallest refractive index that the glass may
have so that the light after entering the rod cannot leave
it through a lateral face irrespective of the values of α?

8.6. At what angle to each other must two flat mirrors
be positioned for a beam of light incident on one of the
mirrors at an arbitrary angle in a plane that is perpendicular to the mirror surface to be reflected from both mirrors in such a manner that the refracted beam is parallel
to the incident beam? Is a prism suitable for this purpose?

8.7. An electric bulb is hanging above the center of a
round table whose radius is R. At what height h must it
be hung for the intensity of illumination at the edge of
the table to be maximal?

8.8. A beam of light propagates through a medium *1* and falls onto another medium, *2*, at an angle α_1. After that it propagates in medium 2 at an angle α_2. The light's wavelength in medium *1* is λ_1. What wavelength has the light in medium 2?

8.9. Two identical coherent sources of light, S_1 and S_2, separated by a distance a produce an interference pattern on a screen. The wavelength of the monochromatic light

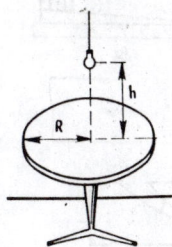 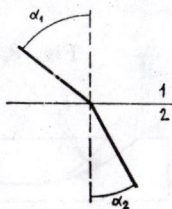 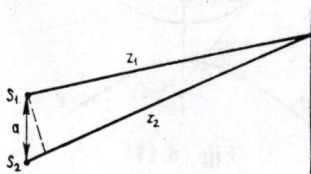

Fig. 8.7   Fig. 8.8   Fig. 8.9

emitted by the sources is λ. Determine the maximal number of interference fringes that can be observed assuming that the screen is infinitely large.

8.10. In an experiment that involves the observation of interference of light via two Fresnel mirrors, the source of light is positioned symmetrically in relation to both mirrors at a distance l from the boundary between them. How does the distance between the first interference fringes on a screen that is positioned far from the mirrors depend on the angle θ between the mirrors?

8.11. When there is interference of light waves emitted by two coherent sources, the geometric locus of points with the same difference in the phases of the oscillations that arrive at that point from the two sources constitutes a surface whose sections with the plane of the drawing are the curves ab and $a'b'$ shown in the figure. What is this surface?

8.12. A transparent dielectric is deposited in the form of a thin film on two substrates made of different dielectrics. Both films form geometrically identical wedge-like layers. The refractive index of the material of the film is n and those of the substrates are n_1 and n_2, with $n_1 < n < n_2$. Suppose that two light beams of similar spectral composition fall on the two systems at the same angle. In what respects do the resulting interference patterns differ?

8.13. An air wedge is illuminated by monochromatic light. The distance between the resulting interference fringes is a. How will the distance between the interference fringes change if the space between the plates that constitutes the wedge is filled with a transparent liquid?

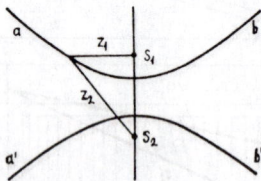

Fig. 8.11

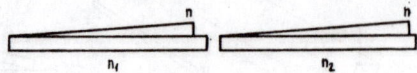

Fig. 8.12

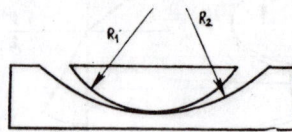

Fig. 8.14

8.14. A plano-convex lens with a radius of curvature R_1 is lying on a reflecting cylindrical surface whose radius of curvature is R_2. The lens is illuminated from above. What shape do the interference fringes have?

8.15. A plano-convex segment of a glass cylinder whose curvature radius is R is lying on a flat plate. A parallel beam of light falls on this segment from above. What shape will the interference fringes have and how will the distance between the fringes change as we move away from the straight line along which the segment touches the plate?

8.16. During observation of Newton rings, a small particle of unknown thickness a got between the lens and the plate. How can one determine the wavelength of monochromatic light incident from above on the lens using only graphical considerations? What scales along the vertical and horizontal axes are preferable?

8.17. On a reflecting substrate there lies a transparent plane-parallel plate that forms an angle α with the substrate. Thus a wedge-like film of air is formed. The substrate has a triangular ledge whose cross section is an isosceles triangle with angles θ at its base. The plate is illuminated with monochromatic light from above. Assuming that the angles α and θ are small, sketch the positions of the interference fringes. The size of the wavelength is shown in the figure.

8.18. In the observation of the interference pattern in an air wedge (Figure (a)) there sometimes appear interference fringes with distortions caused by the presence of a ledge or a dent on the substrate. Which of the two inter-

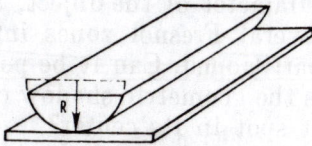

Fig. 8.15

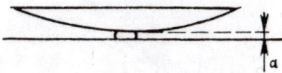

Fig. 8.16

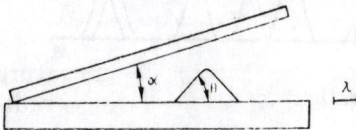

Fig. 8.17

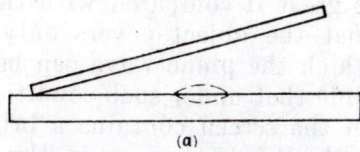

(a)

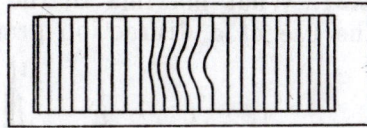

(b)

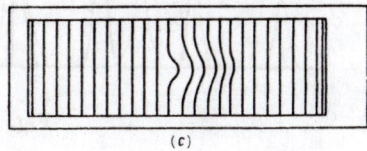

(c)

Fig. 8.18

ference patterns in Figures (b) and (c) corresponds to which defect?

8.19. Light from a distant source falls on a screen with a round hole. At a certain distance from this screen an-

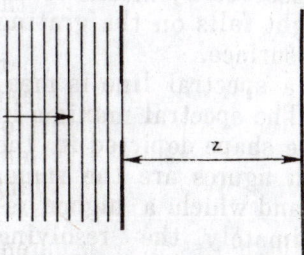

Fig. 8.19

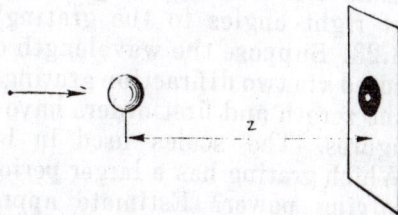

Fig. 8.20

other screen is placed, and it is on this screen that the diffraction pattern is observed. How will the intensity of illumination at the center of the second screen change if the distance between the screens is gradually increased, that is, does the intensity of illumination remain constant or does it monotonically decrease or does it vary periodically?

8.20. Light from a distant monochromatic source, which can be considered point-like, is incident on a small round opaque disk or sphere. A screen is positioned at a certain distance z from the object. This distance, z, is great if compared with the diameter of the object, so that the object covers only several Fresnel zones into which the plane wave can be partitioned. Can it be possible that under such conditions the geometric shadow on on the screen contains a bright spot in its center?

8.21. What maxima in the spectrum obtained through the use of a diffraction grating correspond to the line

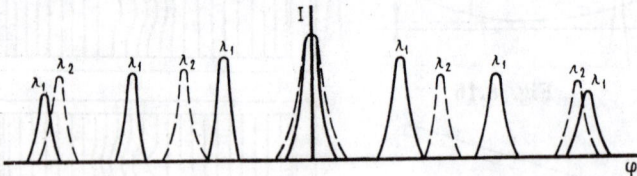

Fig. 8.21

with a longer wavelength and what maxima, to the line with a shorter wavelength? What approximately is the ratio of these wavelengths?

8.22. In a spectrum obtained through the use of a diffraction grating, a spectral line is obtained in the first order at an angle φ_1. Determine the highest order of the spectrum in which this line can be observed by means of the same diffraction grating if the light falls on the grating at right angles to the grating's surface.

8.23. Suppose the wavelength of a spectral line is measured via two diffraction gratings. The spectral maxima in the zeroth and first orders have the shape depicted in the figures. The scales used in both figures are the same. Which grating has a larger period and which, a higher resolving power? Estimate approximately the resolving power of each grating assuming that the natural line width and the Doppler line width are considerably smaller than the one obtained in experiments.

8.24. Suppose there are two diffraction gratings with spacings c_1 and c_2 and a total number of lines N_1 and N_2, respectively. Here $c_1 < c_2$ and $N_1 > N_2$, but the product

cN is the same for both gratings. Which of the two gratings has a greater maximal resolving power if the same spectral line is observed at normal incidence of light on the gratings?

8.25. A parallel beam of light falls at an angle θ on a flat diffraction grating with a spacing d. Determine the fundamental grating condition for the wavelength λ, the maximum order of the spectrum in which the appropriate spectral line can be observed, the maximum wavelength for which a line in the spectrum can be resolved, and the maximum dispersive power of the grating?

8.26. A phonograph record can be used as a reflecting diffraction grating. To obtain a clear diffraction pattern,

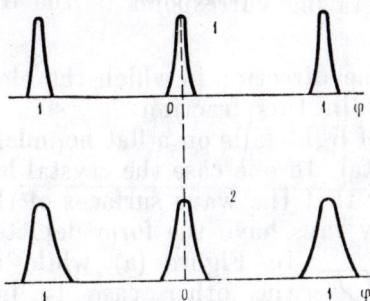

Fig. 8.23

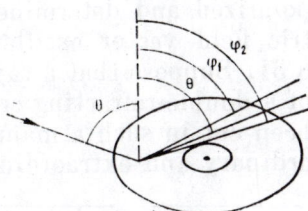

Fig. 8.26

one must direct the light at an angle that is as close to the grating angle to the surface of the record as possible. Why?

8.27. What minimal value can the Brewster angle have when light falls from air onto the surface of any dielectric?

8.28. When light is incident on a transparent dielectric at the Brewster angle ($\tan \alpha = n$), the reflected light proves to be completely polarized. Is the refracted light also completely polarized in this case?

8.29. Natural nonpolarized light is incident on a double-refracting crystal. The normals to the ordinary wave (o) and the extraordinary wave (e) are directed as shown in the figure. Find the ratio of the wavelengths of these waves.

8.30. A T-shaped pipe with blackened walls is filled with a turbid medium. Light falls onto one end of the pipe in the direction designated by *1*. As a result of scattering, a fraction of the light emerges from the pipe in the direction designated by *2*. Prove that this fraction is

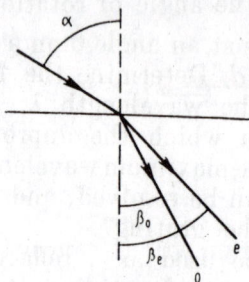

Fig. 8.29

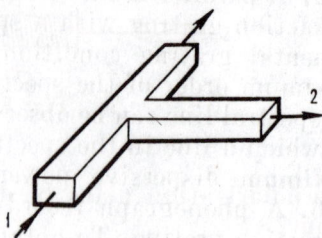

Fig. 8.30

polarized and determine the direction in which the electric field vector oscillates in this fraction.

8.31. Suppose that a ray of light falls on a flat boundary of a double-refracting crystal. In one case the crystal has been cut in such a manner that the wave surfaces of the ordinary and extraordinary rays have the form depicted in Figure (a), while in

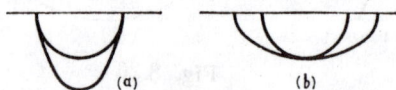

Fig. 8.31

Fig. 8.32

the other case it has been cut in such a manner that the corresponding wave surfaces have the form shown in Figure (b). How is the optic axis of the crystal directed in each case and is the crystal positive or negative?

8.32. Natural light with intensity I_0 passes through two Nicol prisms whose transmission planes are at an angle θ to each other. After the light has passed through the second prism, it falls on a mirror, is reflected by the mirror, and passes through the two Nicol prisms once more. What is the intensity I of the light that has travelled this path?

8.33. Polarized light passes through a transparent substance that is placed in a longitudinal magnetic field. The result is the so-called Faraday effect (rotation of the polarization plane in a magnetic field). After passing through the substance (and magnetic field), the light is reflected by a mirror and travels in the opposite direction, whereby it travels through the magnetic field once more but in the opposite direction. Will the angle of rotation

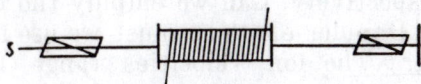

Fig. 8.33

of the polarization plane be doubled or will it cancel itself out?

8.34. When an electric field is applied to a capacitor that is submerged in nitrobenzene, artificial anisotropy emerges in the medium and the nitrobenzene behaves like a

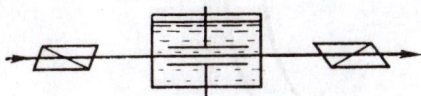

Fig. 8.34

double-refracting crystal in which the reftactive index of the extraordinary ray, n_e, is greater than that of the ordinary wave, n_o. The phenomenon, known as the Kerr effect, can be observed via two crossed Nicol prisms. Does the observed pattern change if the direction of the electric field is reversed?

8.35. When a source of light moves toward the observer, the optical Doppler effect manifests itself. The curves in the figure depict the dependence of the perceptible frequency of the light on the speed of the source of light, with one curve corresponding to the results predicted by classical theory and the other. to the results predicted by the theory of relativity. The ratio of the speed of the source to the speed of light is laid off on the horizontal axis, while the ratio of the perceptible frequency to the frequency of the light emitted by the source (i.e. of a fixed source) is laid off on the vertical axis. Which curve corresponds to which theory?

8.36. To determine the directional velocity of the ions that move in an electric field in a plasma, one commonly

measures the wavelength of the waves emitted by the excited ions. The measurements are carried out in two directions, counter to the direction of motion of the ions and
"in pursuit" of the ions. The measured wavelengths are
λ_1 and λ_2, respectively. Can we employ the classical formulas of the Doppler effect or must we use the relativis-
tic formulas? The ion velocities range from 10^4 to
10^5 m/s.

8.37. The figure depicts the same spectral line emitted
by a gas at different temperatures. The wavelength is
laid off on the horizontal axis, while the ratio of the inten-

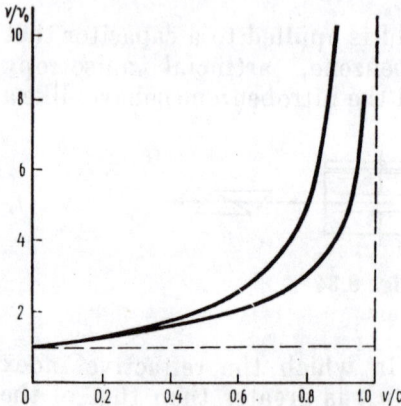

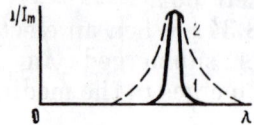

Fig. 8.35   Fig. 8.37

sity at a given wavelength to the maximal intensity at a
given temperature is laid off on the vertical axis. Which
curve corresponds to a higher temperature?

8.38. An electric current flows through a rarefied gas in
a tube *1* (Figure (a)). The radiation emitted by the excited positive ions is analyzed in the transverse direction
by a spectrograph 2. The wavelength distribution of the
intensity of the radiation for one spectral line is shown in
Figure (b). Can analyzing this distribution yield the temperature of the ions?

8.39. Two objects having the same shape and size but
different absorption coefficients (immisivities) are heated to the same temperature and placed in a vacuum.
As a result of emission of radiation the objects cool off.
The curves in the figure show the change in temperature

in the process of cooling. The cooling-off time from the
moment the objects were placed in the vacuum is laid off
on the horizontal axis, while the temperature of the objects is laid off on the vertical axis. Which curve character-

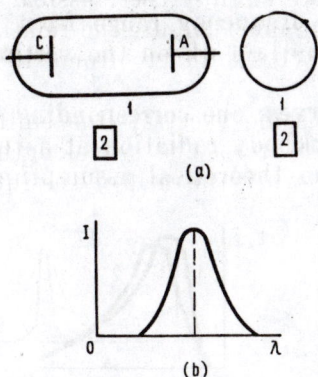

Fig. 8.38

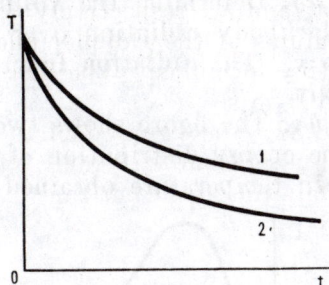

Fig. 8.39

izes the object with a higher absorption coefficient and
which, with a lower absorption coefficient?

8.40. An ideal gas is placed inside a closed isolated volume. The concentration of the molecules of the gas is
n. At what temperature will the volume density of the
kinetic energy of translational molecular motion in the
gas be equal to the volume density of the energy of blackbody (electromagnetic) radiation? Illustrate the result
with numerical examples.

8.41. Two separate segments of equal area are isolated
in the energy distribution of blackbody radiation. Are

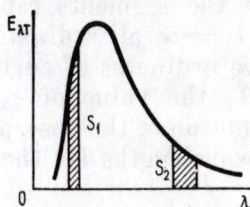

Fig. 8.41

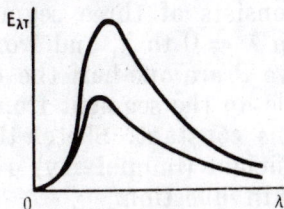

Fig. 8.42

the emissive powers over the respective wavelength intervals the same? What about the number of emitted photons in each segment?

8.42. A student has sketched the curves representing the energy distribution in the emission spectrum of blackbody radiation for two temperatures as shown in the figure. What mistake did the student make?

8.43. Determine the volume density of the energy of blackbody radiation over the frequency range from ν_1 to ν_2. The radiation function is laid off on the vertical axis.

8.44. The figure shows two curves: one corresponding to the energy distribution of blackbody radiation at a certain temperature obtained from theoretical assumptions

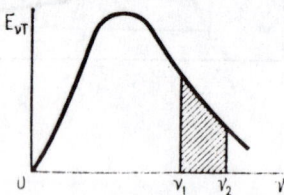

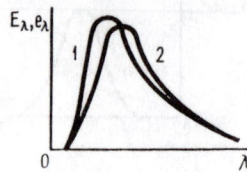

Fig. 8.43   Fig. 8.44

(curve *1*), and the other corresponding to the energy distribution of the radiation emitted by a certain object that has been heated to the same temperature (curve *2*). Why can we be sure that the experimental curve does not give a true picture?

8.45. Curve *1* in the figure depicts the energy distribution in the emission spectrum of a black body. Curve *2* represents, in schematic form, the energy distribution in the emission spectrum of a certain object that has been heated to the same temperature as the black body. Curve *2* consists of three segments: on the segments ranging from $\lambda = 0$ to λ_1 and from λ_2 to $\lambda = \infty$ all ordinates of curve *2* are one-half the respective ordinates of curve *1*, while on the segment from λ_1 to λ_2 the value of e_λ remains constant. Sketch the distribution of the absorption coefficient (immisivity) over the wavelengths for the object in question.

8.46. The radiation emitted by a black body can be represented either by the energy distribution over the wavelengths (Figure (a)) or by the energy distribution over the frequencies (Figure (b)). In the first case the wavelength at which the black body emits a maximum amount of radiation is λ_m, while in the second the frequency at which the black body emits maximum amount of radiation is ν_m. Is it true that at a fixed temperature the quantities λ_m and ν_m are related through the formula $\nu_m = c/\lambda_m$?

8.47. Represent the volume density of the energy of blackbody radiation in the form of a distribution function for the number of quanta in the energy of one quantum.

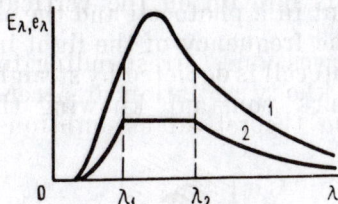

Fig. 8.45

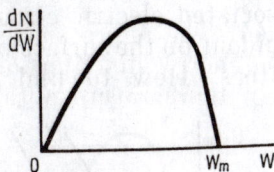

Fig. 8.51

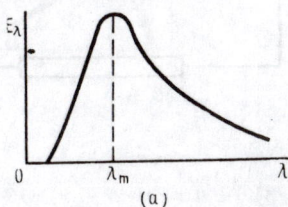

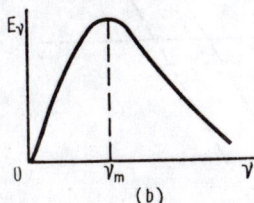

Fig. 8.46

8.48. How does the volume specific heat capacity of the vacuum depend on temperature?

8.49. According to the electromagnetic theory of light, the light incident on a surface always exerts a pressure on that surface equal to

$$p = \frac{I}{c}(1+R), \qquad (8.49.1)$$

where I is the intensity of the light, that is, the light energy arriving every second at a unit area of the surface, and R is the reflection coefficient. Can the origin of this pressure be explained in the same manner as is done in the kinetic theory of gases, where the pressure of a gas on the wall of a vessel is interpreted as transfer of momentum from each particle to the wall?

8.50. Are there any practical means by which one can obtain a beam of parallel rays of light in the mathematical sense (using the terminology of wave optics, a stream of strictly plane waves)?

8.51. The energy distribution function for photoelectrons has the form shown in the figure. What determines the maximal energy of the photoelectrons?

8.52. In the Lukirskii-Prilezhaev experiments (also conducted independently by R. A. Millikan), the dependence of the stopping potential U_{stop}, that is, the potential needed to stop the photocurrent in a photocell and the associated electric circuit, on the frequency of the light incident on the surface of the photocell is depicted by straight lines. How to find the Planck constant knowing the

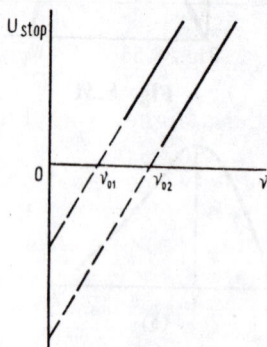

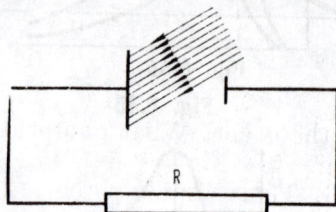

Fig. 8.52   **Fig. 8.53**

slope of these straight lines? In what respect do the parameters that characterize these two straight lines differ?

8.53. Two electrodes placed in a vacuum at a certain distance from each other are connected electrically by a resistor. One electrode is illuminated with light from a source whose spectrum contains radiation with a wavelength λ that satisfies the condition

$$hc/\lambda > p,$$

where p is the work function of electrons leaving the metal of the illuminated electrode. Will there be any current in this circuit?

8.54. A photocathode can be illuminated by the light from two sources, each of which emits monochromatic radiation. The sources are positioned at equal distances from the photocathode. The dependence of the photocurrent on the voltage between the cathode and the anode is depicted by curve *1* for one source and by curve *2* for the other. In what respect do these sources differ?

8.55. Two photocathodes are illuminated by the light emitted by a single source. The dependence of the photocurrent on the voltage between the cathode and the anode is depicted by curve *1* for one cathode and by curve *2*

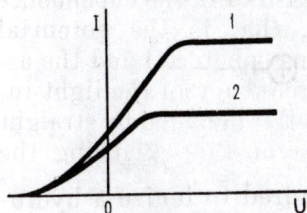

Fig. 8.54

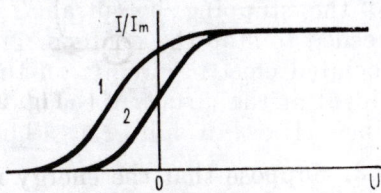

Fig. 8.55

for the other? What photocathode has a higher work function?

8.56. The stopping potential applied between a photocathode and the respective anode is such that the fastest photoelectrons can fly only one-half of the distance between the cathode and the anode. Will the electrons be able to reach the anode if the distance between the cathode and the anode is reduced by half but the voltage is kept constant?

8.57. In one case of Compton scattering a photon flies at an angle θ to the initial direction of the incident photons,

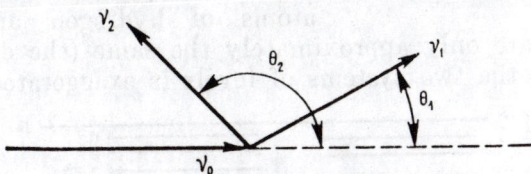

Fig. 8.57

and in other case it flies at an angle θ_2. In which case is the wavelength of the radiation after scattering greater, and in which case does the electron participating in the interaction receive a greater portion of energy?

9. Atomic and Nuclear Physics

9.1. A proton that has flown over a great distance hits a proton that is at rest. The impact parameter is zero, that is, the velocity of the incident proton is directed along the straight line connecting the centers of the protons. The mass of the proton is known, m, and the initial velocity of the incident proton is v_0. How close will the incidence proton get to the fixed proton?

Fig. 9.1

9.2. Suppose that the energy required to ionize a hydrogen atom is W_1. Must the electron, the hydrogen ion, and the helium ion have the same initial kinetic energies for the hydrogen atom to become ionized?

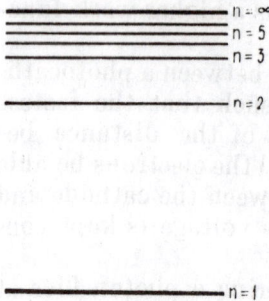

Fig. 9.3

9.3. The system of quantum levels of an atom is assumed to be like the one depicted in the figure. How will each of the energy components of the electron (the kinetic energy and the potential energy) vary if the electron moves from a lower level to a higher level?

9.4. The quantum levels of atoms of hydrogen and deuterium are only approximately the same (the difference between the two systems of levels is exaggerated in the

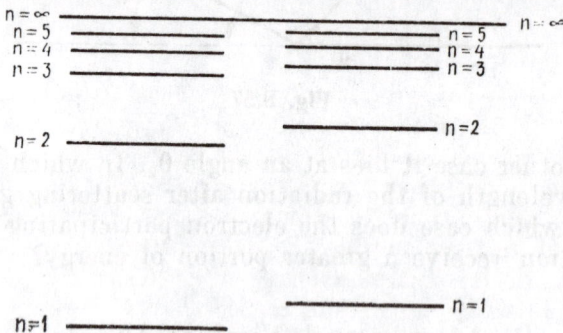

Fig. 9.4

figure). Which system of levels belongs to which atom? What is the reason for this discrepancy?

110

9.5. Every other spectral line in one of the spectral series of an ionized helium atom (the Pickering series) closely resembles a line in the Balmer series for hydrogen. What is the principal quantum number of the level to which the electrons transfer when these lines are emitted?

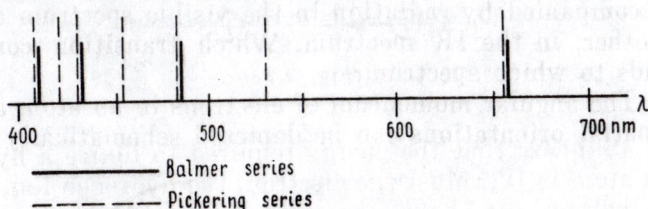

Fig. 9.5

Why don't the lines coincide exactly? What is the meaning of the lines that lie in between the lines of the Balmer series?

9.6. Four lines in the Balmer series lie in the visible part of the spectrum. What must the principal quantum number of the electron level in a doubly ionized lithium atom be for the lines emitted when electrons go over to this level to lie close to the lines of the Balmer series? What is the overall number of lines lying in this wavelength region?

9.7. An electron moving in an atom is acted upon by the Coulomb force of attraction generated by the nucleus. Can an external electric field be created that is capable of neutralizing the Coulomb force and ionizing, say, a hydrogen atom? Field strengths that can be created by modern devices are about 10^7 to 10^8 V/m.

9.8. In a He-Ne laser, the helium atoms are excited from the ground state to two sublevels, 2^1S and 2^3S, interact with Ne atoms, and give off their energy to Ne atoms, with the result that the latter are transferred to the $3S$ and $2S$ levels. The Ne atoms in these states emit radiation and go over to the $2P$ level. In the figure, the $3S$ and $2S$ levels, each consisting of four sublevels, and the $2P$ level, which consists of ten sublevels, are depicted by broad black bands. In addition to the above-mentioned transitions, a transition from the $3S$ state to the $3P$ level is possible, but we do not show this transition in the figure. From the $2P$ state, Ne atoms go over to the $1S$ state, and then gradually return to their ground state.

Why don't He atoms emit radiation during transitions from the 2^1S and 2^3S states directly to the ground state? What must be the relationship between the lifetimes of He atoms in states $3S$, $2S$, and $2P$ for continuous generation of radiation to be possible? It has been established that of the two transitions, $3S \rightarrow 2P$ and $2S \rightarrow 2P$, one is accompanied by radiation in the visible spectrum and the other, in the IR spectrum. Which transition corresponds to which spectrum?

9.9. The angular momentum of electrons in an atom and ts spatial orientations can be depicted schematically by

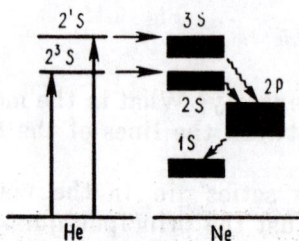

Fig. 9.8

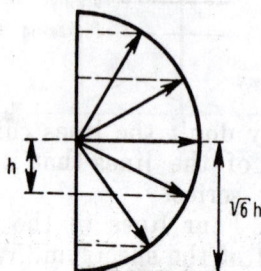

Fig. 9.9

a vector diagram where the length of the vector is proportional to the absolute value of the orbital angular momentum of an electron. What vectors in the diagram correspond to the minimal value of the principal quantum number n and what are the values of the quantum numbers l and m?

9.10. In the Stern-Gerlach experiment, which was conducted with the aim of discovering the spatial quantization of an atomic magnetic moment, a beam of silver atoms is sent through a nonuniform magnetic field generated by magnets whose configuration is shown in the figure. Why does the experiment require a nonuniform field?

9.11. The intensity distribution of X-ray radiation over wavelengths consists of a continuous spectrum, which is limited from the short-wave side by a limit wavelength λ_m, and a characteristic spectrum, which consists of separate peaks. In the figure (with an arbitrary scale) we depict such a distribution for a voltage U_1 applied to the X-ray tube. How will the distribution change if the voltage is decreased three-fold, that is, $U_2 = (1/3) U_1$?

9.12. An electron is inside a potential well with vertical walls. The electronic wave function is depicted in the figure. Is the depth of the well finite or infinite?

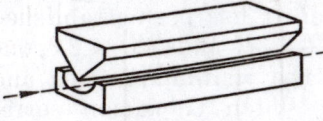

Fig. 9.10

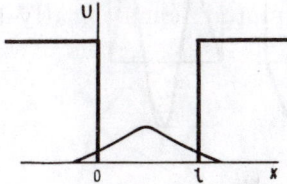

Fig. 9.11

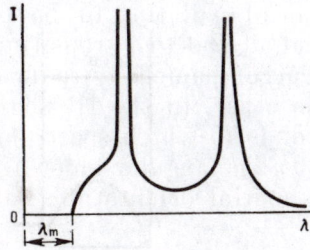

Fig. 9.12

9.13. An electron is in motion in a potential well of infinite depth. Depending on the electron kinetic energy, the electronic wave function has different configurations depicted in the figure. Which of these states is retained when the width of the potential well is decreased two-fold? By what factor will the minimal kinetic energy of the electron change in the process?

9.14. From the viewpoint of the optical analogy of the wave properties of an electron, the regions of space where it possesses different potential energies may be interpreted as regions with different refractive indices. In the figure two such regions are depicted, the regions are separated by a boundary where the potential energy P experiences a jump. In which of these regions is the refractive index greater? In which of the two cases, when the electron moves from left to right or when it moves from right to left, will the phase of the wave function be retained under reflection of the electron from the barrier, and in which will it change to its opposite?

9.15. An electron moving from left to right meets an obstacle, which in one case is a step (Figure (a)), and in the other a barrier (Figure (b)). What are the probabilities of the electron overcoming the step and the barrier according to the classical theory and the quantum theory in two separate cases, namely, when the electron kinetic energy E is lower than P and when it is higher than P?

9.16. An electron moving from left to right passes through three regions: I, II, and III. Its kinetic energy in re-

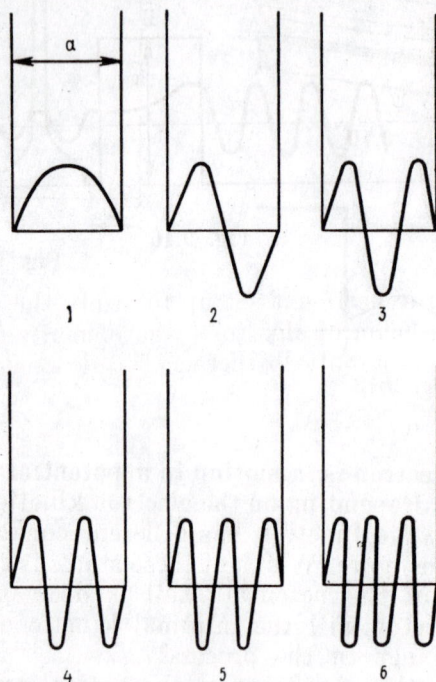

Fig. 9.13

gions I and III is the same, E. Assuming the potential energy in these two regions to be zero, find the relationship between the kinetic energy E and the potential

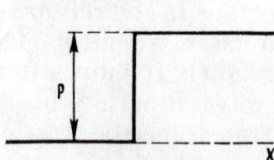

Fig. 9.14

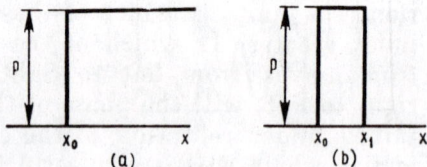

Fig. 9.15

energy P in region II if the electronic wave function has the configuration depicted in the figure.

9.17. According to classical kinetic theory, absolute zero is the temperature at which molecular motion ceases.

In relation to a solid body, this means that the thermal oscillatory motion of atoms or molecules forming the crystal lattice also ceases. Is the same conclusion valid from the standpoint of quantum mechanics?

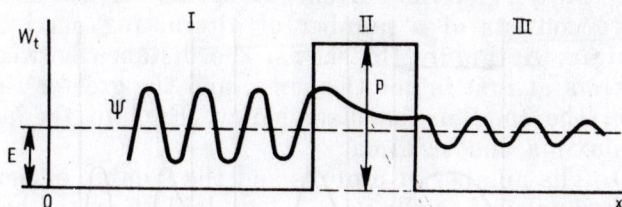

Fig. 9.16

9.18. In an experiment set up to study the diffraction of electrons, a beam of electrons whose energy can be varied by varying a potential difference is directed to a surface

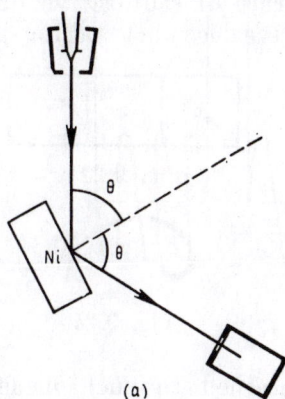

(a)

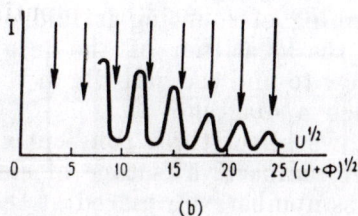

(b)

Fig. 9.18

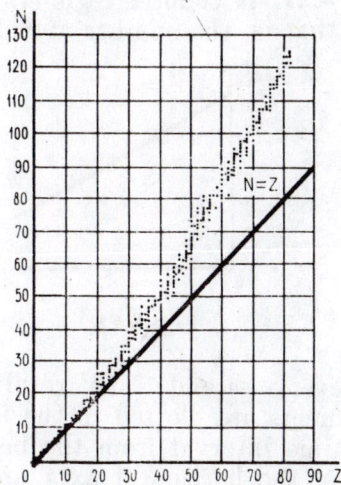

Fig. 9.19

of a single crystal at an angle θ. The diffracted (scattered) beam is analyzed by a detector positioned at the same angle θ (Figure (a)). In the experiment, the current of the scattered electrons was measured as a function of the applied potential difference used to accelerate the electrons. The results were plotted on a diagram, with the square root of the accelerating voltage laid off on the horizontal axis and the electron current, along the vertical axis. The curve consists of a number of alternating maxima and minima. As Figure (b) shows, the distance between the maxima at first is not the same, and the greater the voltage, the smaller the discrepancy. Explain the pattern of maxima and minima.

9.19. The number of protons and the number of neutrons in the nuclei of stable isotopes are laid off on the horizontal and vertical axes, respectively. Why does the fraction of neutrons in the overall number of nucleons increase with the mass number of the nuclei?

9.20. How many nucleons can there be in a nucleus on the lowest quantum level?

9.21. A counter registers the rate of radioactive decay, that is, the number of radioactive decay acts taking place

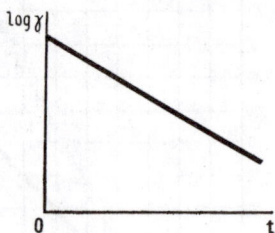

Fig. 9.21

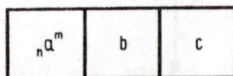

Fig. 9.22

Fig. 9.24

every second. The results obtained in such measurements are plotted in the form of a diagram in which the time interval from the beginning of counting is laid off on the horizontal axis and the logarithm of the decay rate, on the vertical axis. How to find the half-life of the radioactive element from such a diagram?

9.22. In the Periodic Table, we select three consecutive elements, say, a, b, and c. A radioactive isotope of element a whose proton and mass numbers are placed at the symbol of the element transforms into element b, which in turn transforms into element c. This last element transforms into an isotope of the initial element a. What processes cause the transformations $a \to b$, $b \to c$, and $c \to a$? What are the proton and mass numbers of the nuclei of elements b and c and those of the nucleus of element a after the final transformation is completed?

9.23. A number N_0 of atoms of a radioactive element are placed inside a closed volume. The radioactive decay constant for the nuclei of this element is λ_1. The daughter nuclei that form as a result of the decay process are assumed to be radioactive, too, with a radioactive decay constant λ_2. Determine the time variation of the number of such nuclei. Consider two limiting cases: $\lambda_1 \gg \lambda_2$ and $\lambda_1 \ll \lambda_2$.

9.24. The track of a beta particle (an electron) in a Wilson chamber has the shape of a limacon (a spiral). Where does the track begin and where does it end? How is the magnetic field that forces the beta particle to move in this manner directed?

9.25. In beta decay, the velocity of the nucleus that emits an electron is not directed along the line along which the electron velocity is directed. How can this phenomenon be explained?

9.26. The track of a proton in a Wilson chamber has a "knee", where the proton changes its trajectory by $45°$.

Fig. 9.25   Fig. 9.26   Fig. 9.27

Momentum and energy conservation implies that the proton has collided with a neutron. Which of the two particles has a higher energy if the neutron is considered to be initially at rest and free?

9.27. The track of an alpha particle in a Wilson chamber filled with a gas has a "knee", where the particle changes its direction of flight by an angle greater than $90°$. Starting with what gas in the Periodic Table is such a track possible?

9.28. Two radioactive ions are emitted by an accelerator in the same direction with the same velocity $\mathbf{v}$ whose absolute value is close to the speed of light. Following this event, the nuclei of the ions emit electrons (each nucleus emits one electron). The velocity of one electron coincides in direction with **v** while the velocity of the other electron is in opposition to **v**. With respect to the nuclei the electron velocities (their absolute values, that is) are the

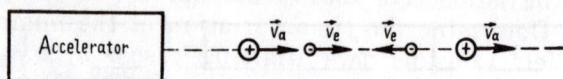

Fig. 9.28

same, v. Find the electron velocities with respect to the (fixed) accelerator and the velocity of one electron with respect to the other.

9.29. Within the framework of the "classical" Bohr theory, an excited atom is an atom one electron of which moves along an orbit that is farther from the nucleus

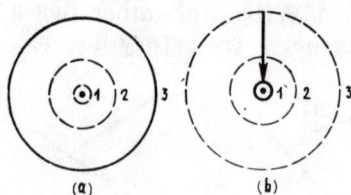

Fig. 9.29

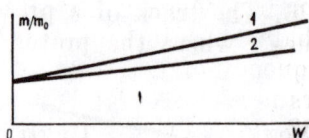

Fig. 9.30

than in the ground state (Figure (a)). When the atom goes over to its ground state (Figure (b)), the atom emits a photon. In the literature, especially popular-science literature, the common way to describe this process is to say that mass has transformed into energy. Is this actually the case?

9.30. Two charged particles acquire equal energy when moving in an accelerator. The dependence of the mass of each particle on the energy acquired is depicted by curves *1* and *2* in the figure. Which of the two particles has a greater rest mass?

9.31. The principle of operation of a linear accelerator is illustrated in the figure accompanying the problem. A charged particle is emitted by a source and is accelerated by a potential difference U between source S and cylinder *1*. During the time it takes the particle to fly through cylinder *1*, the potential difference between *1* and the next cylinder, *2*, changes its sign and, leaving cylinder *2*, the particle again finds itself in an accelerating field with the same potential difference U. The length of cylinder *2* is selected such that when the particle leaves this cylinder, the field will again change sign, so that the particle is accelerated anew, and so on. If the particle has

Fig. 9.31

passed N gaps between the cylinders, the energy it acquires is $W = eUN$ (it is assumed that the particle is singly charged). Since as the particle is accelerated the path it traverses in the course of a single change in polarity between the cylinders increases, each subsequent cylinder must be longer than the previous one. However, at a certain high energy the size of cylinders ceases to grow. What determines the maximal length of a cylinder if the frequency of variation of the voltage between the cylinders is ν?

9.32. Why cyclotrons are not employed to accelerate electrons? What generated a need for building more complex accelerators such as the synchrocyclotron and the synchrophasotron?

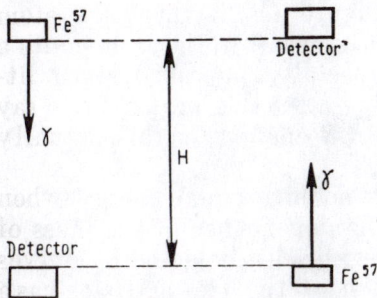

Fig. 9.33

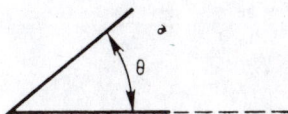

Fig. 9.34

9.33. Two samples of radioactive iron ^{57}Fe emit gamma-ray quanta. One sample is placed at an altitude H above sea level and the appropriate detector at sea level, while the second sample is placed at sea level and the appropriate detector at altitude H. Which of the two detected quanta has a higher frequency?

9.34. In observing Cerenkov radiation it was found that light propagates at an angle θ to the direction of electron motion. Find the refractive index of the substance in which the radiation is excited.

Answers and Solutions

1. Fundamentals of Mechanics

1.1. If $AB = AC = l$, then the times of flight from A to B and from B to A are, respectively, $l/(c - v)$ and $l/(c + v)$. The entire flight time is

$$t_1 = \frac{l}{c-v} + \frac{l}{c+v} = \frac{2lc}{c^2 - v^2}.$$

For the second airplane to fly from A to C, its velocity must be directed at an angle to the direction of the wind

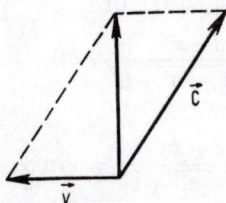 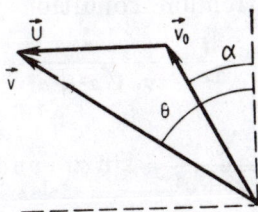

Fig. 1.1　　　　　　　　**Fig. 1.2**

in such a manner that the resulting velocity directed toward C is equal to $(c^2 - v^2)^{1/2}$ in magnitude. The entire flight time of this airplane will be

$$t_2 = \frac{2l}{\sqrt{c^2 - v^2}}.$$

The second airplane will arrive before the first, and the flight time ratio is

$$t_2/t_1 = \sqrt{1 - v^2/c^2}.$$

1.2. The figure shows that

$$\tan \theta = \frac{v_0 \sin \alpha + u}{v_0 \cos \alpha} = \tan \alpha + \frac{u}{v_0 \cos \alpha}.$$

Velocity v can be found from the equation

$$(v_0 \sin \alpha + u)^2 + v_0^2 \cos^2 \alpha = v^2,$$

which yields

$$v = v_0 \sqrt{1 + 2\frac{u}{v_0}\sin\alpha + \left(\frac{u}{v_0}\right)^2}.$$

The boat will travel directly across the river if $\theta = 0$. Under this condition, $\sin\alpha = -u/v_0$. Obviously, the boat can travel at right angles to the current only if v_0 is greater than u.

1.3. The time of travel by boat from A to C is

$$t_1 = \sqrt{x^2 + a^2}/v_1.$$

The time of travel by foot from C to B is

$$t_2 = \sqrt{(d-x)^2 + b^2}/v_2.$$

The total time of travel is

$$t = t_1 + t_2 = \frac{\sqrt{x^2 + a^2}}{v_1} + \frac{\sqrt{(d-x)^2 + b^2}}{v_2}.$$

The extremum condition is $dt/dx = 0$, or

$$\frac{dt}{dx} = \frac{x}{v_1 \sqrt{x^2 + a^2}} - \frac{d-x}{v_2 \sqrt{(d-x)^2 + b^2}} = 0.$$

Since

$$\frac{x}{\sqrt{x^2 + a^2}} = \sin\alpha_1 \quad \text{and} \quad \frac{d-x}{\sqrt{(d-x)^2 + b^2}} = \sin\alpha_2,$$

we can write $\sin\alpha_1/v_1 = \sin\alpha_2/v_2$, whence

$$\frac{\sin\alpha_1}{\sin\alpha_2} = \frac{v_1}{v_2}.$$

We can easily see that the extremum corresponds to the minimum of time of travel.

1.4. The time of travel along straight line BC is determined by the length S of segment BC and the acceleration w. The figure shows that

$$S = \sqrt{a^2 + h^2}, \quad w = \frac{h}{\sqrt{a^2 + h^2}}\, g.$$

Since $S = wt^2/2$, we can write

$$\sqrt{a^2 + h^2} = \frac{g}{2}\frac{h}{\sqrt{a^2 + h^2}}\, t^2,$$

whence

$$t = \sqrt{\frac{2}{g}\frac{(a^2 + h^2)}{h}}.$$

Nullifying the derivative (the extremum condition),

$$\frac{dt}{dh} = \frac{h^2 - a^2}{\sqrt{2gh^3(a^2 + h^2)}} = 0,$$

yields $h = a$.

The same result is obtained if we express S and w in terms of α:

$$S = a/\cos\alpha, \quad w = g\sin\alpha,$$

$$t = \sqrt{\frac{2}{g} \frac{a}{\sin\alpha \cdot \cos\alpha}}.$$

Nullifying the derivative $dt/d\alpha$, we find that $\alpha = 45°$.

1.5. The acceleration in rectilinear motion is the second derivative of the distance traveled with respect to time. For a concave curve the second derivative is positive, while for a convex curve the second derivative is negative, whereby curve (a) corresponds to decelerated motion and curve (b) to accelerated motion.

1.6. By definition, acceleration is the time derivative of velocity, $\mathbf{w} = d\mathbf{v}/dt$. For rectilinear motion the vector equation can be written in scalar form. The acceleration is the highest when the derivative is the greatest, that is, when the curvature of the curve is maximal. The curvature is determined by the slope of the tangent line to the particular point on the curve. This corresponds to moment 2 on the time axis. Note that for curvilinear motion the question contains an ambiguity, since to determine the acceleration we must know the radius of the trajectory at every moment in the course of the motion in addition to the magnitude of the velocity. To find the average velocity, we must know the distance traveled by the particle in the course of a definite time interval. In terms of the velocity vs. time graph, the distance traveled is the area of the figure bounded by the curve, the time axis, and the vertical straight lines passing through the initial and final moments of time on the time axis. Analytically the distance is calculated via the integral

$$S = \int_{t_1}^{t_2} v\, dt,$$

whence the average velocity is

$$v = \frac{\int_{t_1}^{t_2} v\, dt}{t_2 - t_1}.$$

1.7. In terms of the velocity vs. time graph, the distance traveled is determined by the area bounded by the curve and the time axis. This area is

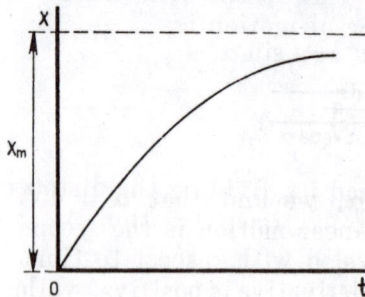

Fig. 1.8

$$S = \frac{\pi}{4}\, v_{\mathrm{m}} t.$$

The average velocity is

$$v = \frac{S}{t} = \frac{\pi}{4}\, v_{\mathrm{m}}.$$

Such motion cannot be realized in practical terms since at the initial and final moments of the motion the acceleration, which is dv/dt, is infinitely large in absolute value.

1.8. The particle will never get to point B but will approach it without bound. Indeed, from the equation $v = v_0 - ax$ we get

$$\frac{dx}{v_0 - ax} = dt.$$

Integration of this expression yields

$$\ln\left(\frac{x - v_0/a}{-v_0/a}\right) = -at,$$

whence

$$x = \frac{v_0}{a}(1 - e^{-at}). \qquad (1.8.1)$$

The limit value $x_{\mathrm{m}} = v_0/a$ can be attained only at $t \to \infty$. The dependence of x on t defined by Eq. (1.8.1) is represented by the curve shown in the figure.

1.9. The acceleration

$$w = \frac{dv}{dt} = \frac{dv}{dx}\frac{dx}{dt} = k(v_0 + kx)$$

increases with x. The same result can be obtained from the following line of reasoning: at constant acceleration the relationship between the velocity and the distance traveled is given by the formula

$$v^2 = v_0^2 + 2wx,$$

so that the velocity increases in proportion to the square root of the distance. Hence, for the velocity to increase linearly with x, the acceleration must increase.

1.10. The train covers the distance dx in the course of $dt = dx/v(x)$, where $v(x)$ is the speed with which it travels over dx. The total time of motion is

$$t = \int\limits_0^S \frac{dx}{v(x)}.$$

The average speed is determined by dividing the distance covered by the train by the entire time of motion:

$$v_{av} = \frac{S}{\int\limits_0^S \dfrac{dx}{v(x)}}.$$

If the graph cannot be represented by a formula, it can be reconstructed into the $1/v$ vs. x graph. In this case the integral in the denominator of the expression for v_{av} can be evaluated by graphical means.

1.11. The speed with which the lower end of the rod moves, $v_x = dx/dt$, can be written in the form

$$v_x = \frac{dy}{dt}\frac{dx}{dy}.$$

Since $x = \sqrt{l^2 - y^2}$, we can write

$$\frac{dx}{dy} = -\frac{y}{\sqrt{l^2 - y^2}},$$

whence

$$v_x = -\frac{y}{\sqrt{l^2 - y^2}}\frac{dy}{dt} = \frac{y|v_y|}{\sqrt{l^2 - y^2}}.$$

Thus, the speed of the lower end gets smaller and smaller and vanishes at $y = 0$.

1.12. Since the drag is proportional to the velocity of the object, so is the acceleration caused by this force (with a

125

minus sign). Hence, by Newton's second law,

$$\frac{dv}{dt} = -g - rv,$$

where r is the proportionality factor. Whence

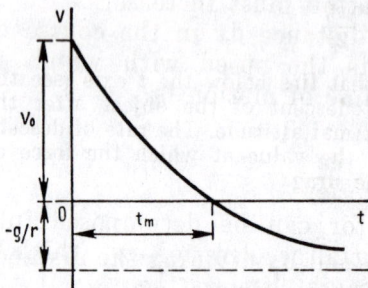

Fig. 1.12

$$\int_{v_0}^{v} \frac{dv}{v + g/r} = -r \int_{0}^{t} dt.$$

Integration yields*

$$v = \left(v_0 + \frac{g}{r}\right) e^{-rt} - \frac{g}{r}. \tag{1.12.1}$$

For $v = 0$ this yields

$$t_{\mathrm{m}} = \frac{1}{r} \ln \left(1 + \frac{rv_0}{g}\right). \tag{1.12.2}$$

To find the maximal altitude, we rewrite (1.12.1) in the form

$$\frac{dh}{dt} = \left(v_0 + \frac{g}{r}\right) e^{-rt} - \frac{g}{r}. \tag{1.12.3}$$

Integrating this equation up to t, we find that

$$h = \left(v_0 + \frac{g}{r}\right) \frac{1}{r} (1 - e^{-rt}) - \frac{g}{r} t. \tag{1.12.4}$$

Bearing in mind that at the point of greatest ascent $v = dh/dt = 0$ and combining this result with (1.12.3), we get

$$\left(v_0 + \frac{g}{r}\right) e^{-rt_{\mathrm{m}}} = \frac{g}{r}. \tag{1.12.5}$$

Combining (1.12.4) with (1.12.5) yields

$$h = \frac{v_0 - gt_{\mathrm{m}}}{r}.$$

Substituting t_{m} from (1.12.2), we arrive at the final result

$$h = \frac{1}{r} \left[v_0 - \frac{g}{r} \ln \left(1 - \frac{rv_0}{g}\right)\right].$$

126

When drag is extremely low, or $rv_0/g \ll 1$, we can employ the expansion

$$\ln\left(1+\frac{rv_0}{g}\right) \approx \frac{rv_0}{g} - \frac{1}{2}\left(\frac{rv_0}{g}\right)^2.$$

This results in the well-known formula

$$h = \frac{v_0^2}{2g}.$$

* The section of the curve that lies below the t axis (see the figure) corresponds to the descent of the object after the object has reached the maximal altitude. The rate of descent asymptotically approaches the value at which the force of gravity is balanced by the drag.

1.13. The acceleration vector can be decomposed into two components, the tangential acceleration w_t, which is directed along the same straight line as the velocity of the particle, and the normal acceleration w_n, which is perpendicular to the velocity. For instance, for $\theta > 90°$ (see Figure (a) accompanying the problem) the tangential acceleration is directed opposite to the particle's velocity and the motion in this case is decelerated, $w < 0$. The presence of a nonzero normal acceleration suggests that the motion is curvilinear. The situation for the other cases is as follows: for $\theta < 90°$ (Figure (b)) the motion is curvilinear and accelerated, for $\theta = 90°$ (Figure (c)) the motion is curvilinear and uniform, and for $\theta = 180°$ (Figure (d)) the motion is rectilinear and decelerated, $w < 0$. Of course, characterizing the motion by the angle between the velocity **v** and the acceleration **w** is meaningful only for a definite moment in time. Subsequent motion may change this characteristic.

1.14. The normal acceleration is

$$w_n = v^2/R = \omega^2 R,$$

whence the linear velocity grows in proportion to the square root of the curvature radius of the spiral, while the angular velocity decreases by the same law.

1.15. When the angle between the total acceleration and the radius becomes equal to $45°$, the normal acceleration becomes equal to the tangential acceleration. Since $w_n = \omega^2 R$ and $w_t = \varepsilon R$, we have $\omega^2 = \varepsilon$, and since $\omega = \varepsilon t$, we have $\varepsilon^2 t^2 = \varepsilon$, with the result that

$$\varepsilon = 1/t^2.$$

1.16. The acceleration with which the object moves is the acceleration of gravity, which at all points of the trajectory is directed vertically downward. From the figure that accompanies the problem we see that as the object ascends the tangential acceleration decreases while the normal acceleration grows. At the highest possible point the tangential acceleration is zero while the normal acceleration is equal to the acceleration of gravity.

1.17. Since at point A the sled's velocity is zero, so is the normal acceleration $w_n = v^2/R$. The tangential accelera-

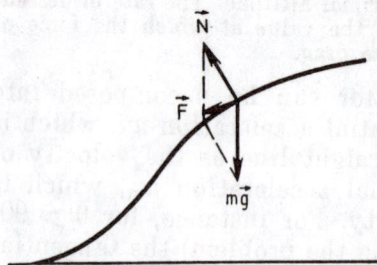

Fig. 1.17

tion is directed down the hill along the tangent to the surface of the hill. The figure accompanying the answer shows the forces that act on the sled. These are the force of gravity mg and the reaction force $\mathbf{N}$ exerted by the surface of the hill. The resultant $\mathbf{F}$ is directed downward along the hill. According to Newton's second law, the acceleration vector points in the same direction as the resultant. If there is friction, the resultant vector does not change direction but becomes somewhat shorter, with the result that the tangential acceleration becomes smaller, too.

1.18. The acceleration vector points in the direction of the resultant of the forces acting on the object. At the lowest possible point only the force of gravity and the reaction force act on the body, provided that there is no friction. This means that at this point the object experiences no tangential acceleration. Since the object is moving along a curvilinear trajectory with a certain velocity, there is a normal acceleration, which is directed toward the center of curvature of the trajectory. This acceleration is generated by the difference between the reaction force exerted by the surface and the force of gravity.

1.19. In the course of time Δt the angular velocity vector will vary from $\boldsymbol{\omega}_1$ to $\boldsymbol{\omega}_2$ without changing its length. The direction of the vector will change by an angle of $\Delta\varphi$. This angle is equal, on the one hand, to $|\Delta\omega|/\omega$ and, on the other, to $\Delta S/R$, where ΔS stands for the displacement of the center of the wheel.* This displacement is equal to $\Omega R \Delta t$, where Ω is the angular velocity of the center of the wheel. Thus,

$$|\Delta\omega|/\omega = \Omega R \Delta t/R \quad \text{and} \quad \varepsilon = \lim_{\Delta t \to 0} \frac{\Delta\omega}{\Delta t} = \omega\Omega.$$

When the wheel is rotating, the point at which it touches the arena will shift in the course of Δt by a distance of $r\omega\Delta t$ on the wheel and by $R\Omega\Delta t$ on the arena. Hence, ω and Ω are linked by the following formula: $\omega r = \Omega R$, whence

$$\varepsilon = \omega^2 \frac{r}{R}.$$

* It is assumed that $\Delta\varphi \ll 1$ rad.

1.20. The height of the center of mass of the vessel with the liquid is determined by the formula

$$h_c = \frac{M(H/2) + m(x/2)}{M+m}, \quad (1.20.1)$$

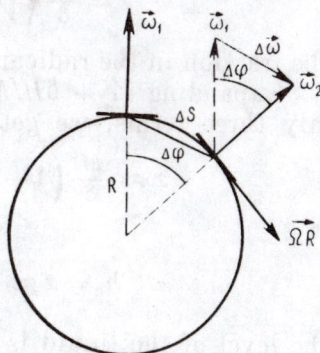

Fig. 1.19

where m is the mass of the liquid. We rewrite (1.20.1) by replacing the mass of the liquid with δx:

$$h_c = \frac{1}{2} \frac{MH + \delta x^2}{M + \delta x}. \quad (1.20.2)$$

Nullifying the derivative of h with respect to x,

$$\frac{dh_c}{dx} = \frac{1}{2} \frac{2\delta x(M+\delta x) - \delta(MH + \delta x^2)}{(M+\delta x)^2} = 0,$$

we get

$$x = \pm \sqrt{\frac{M^2}{\delta^2} + \frac{MH}{\delta}} - \frac{M}{\delta}. \quad (1.20.3)$$

Of course, only the positive value of the root has physical meaning. Substituting this value into (1.20.2), we will find the position of the center of mass. After elementary transformations we get

$$h_c = \sqrt{\frac{M^2}{\delta^2} + \frac{MH}{\delta}} - \frac{M}{\delta}.$$

We have found that the position of the center of mass coincides with the level of the liquid.

Here are some particular cases:

(1) $\delta H = M$ (the liquid filling the vessel completely has a mass equal to the mass of the vessel). Then

$$h_c = x = H \left(\sqrt{2} - 1 \right) \approx 0.41 H.$$

(2) $\delta H \ll M$. Let us transform (1.20.3) to the form

$$x = \frac{M}{\delta} \left(\sqrt{1 + \frac{\delta H}{M}} - 1 \right).$$

The fraction in the radicand is considerably less than unity. Expanding $(1 + \delta H/M)^{1/2}$ in a series and retaining only three terms, we get

$$x \approx \frac{M}{\delta} \left(1 + \frac{\delta H}{2M} - \frac{\delta^2 H^2}{8M^2} - 1 \right),$$

or

$$h_c = x \approx \frac{H}{2} \left(1 - \frac{\delta H}{4M} \right).$$

The level of the liquid is below the middle of the vessel by an insignificant distance.

(3) $\delta H \gg M$. Let us transform (1.20.3) to the form

$$x = H \left(\sqrt{\frac{M^2}{\delta^2 H^2} + \frac{M}{\delta H}} - \frac{M}{\delta H} \right).$$

Bearing in mind that $(M/\delta H)^{1/2} \gg M/\delta H$, we can assume that the expression inside the parentheses in the above formula is simply $(M/\delta H)^{1/2}$, whence

$$h_c = x \approx H (M/\delta H)^{1/2}.$$

The level of the liquid is above the bottom of the vessel by an insignificant distance.

1.21. For the object to be in a state of equilibrium in relation to the wall of the funnel the resultant of the forces acting on the object must impart an acceleration to the object together with the funnel. These forces are the force of gravity and the reaction force exerted by the funnel. Since the force of gravity is constant in this problem and the resultant must be directed horizontally, the direction and magnitude of the reaction force are determined uniquely. But the latter has a different value at different distances from the funnel axis. At a constant angular velocity of the funnel, the greater the radius of rotation the greater the reaction force. For this reason (see the figure accompanying the answer), as the object moves farther from the funnel axis, the resultant of the force of gravity and the reaction force acquires a component directed upward, while as the object moves closer to the axis, the resultant acquires a component directed downward. In

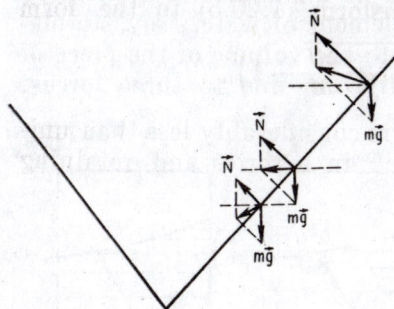

Fg. 1.21

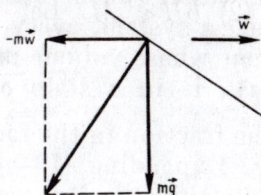

Fig. 1.22

the first case the object tends to move away from the axis still further and rises, while in the second case it tends to move toward the axis and lowers. Thus, the state of equilibrium is unstable.

1.22. It is convenient to think of the vessel with water as a noninertial system. In this case, on each particle of water there acts, in addition to the force of gravity, a force of inertia equal to the product of the particle's mass by the acceleration taken with the minus sign. The surface of water is a plane perpendicular to the vector of the resultant of these two forces. The slope of this surface in relation to the horizontal plane is

$$\tan \alpha = w/g.$$

1.23. Just like in the answer to the previous problem, we can assume the vessel with the liquid to be a noninertial system, in which a force of inertia equal to $-mw = -m\omega^2 x$ acts on every particle of mass m. The resultant of this force and the force of gravity is perpendicular to the surface of the liquid. The derivative dy/dx, equal to the slope of the line tangent to the surface at a given point, is

$$\frac{dy}{dx} = \tan \alpha = \frac{m\omega^2 x}{mg}.$$

Integrating, we find that

$$y = \frac{\omega^2}{2g} x^2.$$

The surface of the liquid is shaped in the form of a paraboloid of revolution.

1.24. Just like in the answers to Problems 1.22 and 1.23, the vessel can be assumed to be a noninertial system. In such a system, every mass element of water, say, an element whose volume is equal to the volume of the piece of cork, is in a state of equilibrium due to three forces:

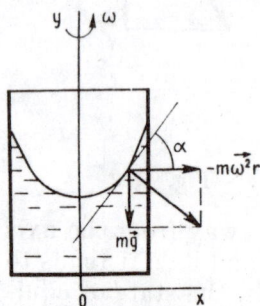

Fig. 1.23

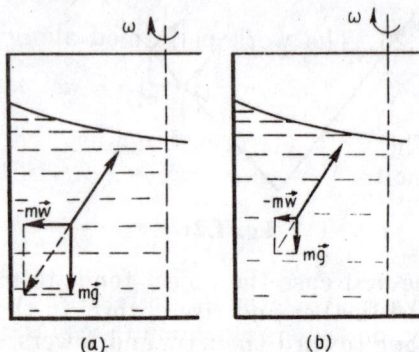

Fig. 1.24

the force of pressure of the surrounding water, the force of gravity, and the force of inertia, which is equal to the product of the element's mass by the normal acceleration of that element taken with the minus sign (Figure (a)). There are also three forces acting on the piece of cork that replaces the element of water: the force of pressure of the surrounding water is the same but the forces of gravity and inertia are lower. As Figure (b) shows, the net force (the difference between the force of pressure and the forces of gravity and inertia) make the cork rise to the surface and, at the same time, move toward the axis of the vessel.

A similar line of reasoning forces us to conclude that an object with a density greater than the density of water, when immersed into a rotating vessel with water, will sink and, in the process, move toward the wall of the vessel.

1.25. According to Newton's second law,

$$\int_0^t F \, dt = \Delta mv.$$

In the case at hand,

$$\int_0^t F \, dt = F_m t/2,$$

whence

$$v = F_m t/2m.$$

1.26. The work performed along ac' is

$$A_1 = ac' \, mgk.$$

The work performed against the forces of friction on the inclined segment ac is

$$A_2 = ac \, mgk \cos \alpha = \frac{ac'}{\cos \alpha} \, mgk \cos \alpha = ac' mgk.$$

We see that the two quantities coincide, and so, obviously, do the similar quantities for $c'b$ and cb. The change in the potential energy about $ac'b$ and acb is zero. Thus, the work performed against the forces of friction along $ac'b$ and that performed against the forces of friction along acb coincide.

1.27. The initial potential energy of the object with respect to the bottom of the hill, mgh, has been used up for work against the force of friction. In returning the body to its initial position, the force performs the same work and, in addition, imparts to the object the initial potential energy. As a result, the total work will be $2mgh$.

1.28. The work performed on an elementary segment of displacement is equal to the decrease in potential energy:

$$dA = -dW.$$

The same work can be represented as the product of force by displacement:

$$dA = F_x dx.$$

Hence

$$F_x = -\frac{dW}{dx} = -2ax.$$

Forces known as quasielastic also obey this law.
1.29. When the object is immersed in the liquid, two forces act on it: the force of gravity and Archimedes' force. If V is the volume of the object, the resultant of these two forces is

$$F = V (\rho_{ob} - \rho_{liq})g.$$

For $\rho_{ob} > \rho_{liq}$, as the object is immersed in the liquid, its potential energy continues to fall below zero, but slower than it would in air. The rate of this decrease is the higher the greater the value of ρ_{ob}. Straight line *1* in the figure accompanying the problem corresponds to an object sinking in a liquid. When $\rho_{ob} = \rho_{liq}$, the potential energy remains constant (straight line *2* coinciding with the x axis). If $\rho_{ob} < \rho_{liq}$, the potential energy of the object begins to increase when the object sinks into the liquid (straight lines *3*, *4*, *5*), and the rate of this increase is the higher the lower the value of ρ_{ob}. The potential energy, while growing, cannot exceed the initial potential energy of the object in air (the dashed horizontal line), and the object can attain this level only when the medium exerts no drag on it. If this is the case, the object will sink to a certain level in the liquid, stop, and then return to the surface with the same speed at the surface as it had when it entered the liquid. Once out of the liquid, the object will rise to the height determined by the initial potential energy. After this it drops back into the liquid, and so on. Of course, under real conditions the drag exerted by the medium will slow down the object, and the greater the viscosity of the liquid the faster this happens.

If the density of the material of the object is one-half the density of the liquid, $\rho_{ob} = (1/2)\rho_{liq}$, then·

$$F = V\rho_{ob}g.$$

In this case the difference between Archimedes' force and the force of gravity is equal (in absolute value) to the latter but is directed in opposition to the force of gravity. The slope of the straight line must be the same as that of the straight line that represents the variation of the potential energy of a falling object. Straight line *4* has such a slope.

134

1.30. The formula that links the force acting on an object with the potential energy of the object,

$$F_r = -\frac{dW}{dr},$$

shows that equilibrium, which occurs when the force is zero, sets in when $dW/dr = 0$. There are two such points on the curve, point *2* and point *4*. Since when the object moves away from point *2* its potential energy increases while when it moves away from point *4* its potential energy decreases, at point *2* equilibrium is stable and at point *4* it is unstable. The fact that a system always tends to a state in which its potential energy is minimal implies that repulsive forces act on the *1-2* and *4-5* segments and an attractive force acts on the *4-2* segment.

1.31. Momentum conservation for the given problem can be written thus:

$$m_1 v_0 = m_1 u_1 + m_2 u_2, \tag{1.31.1}$$

where m_1 is the bullet's mass, m_2 the load's mass, v_0 the initial velocity of the bullet, u_1 the final velocity of the bullet, and u_2 the velocity acquired by the load as a result of the collision. From (1.31.1) it follows that

$$u_2 = \frac{m_1(v_0 - u_1)}{m_2}. \tag{1.31.2}$$

If the bullet flies through the load, after it has left the load it has a velocity that is surely greater than u_2. We write $u_1 = u_2 + V$. Substituting this expression into (1.31.1), we get

$$u_2 = \frac{m_1(v_0 - V)}{m_1 + m_2}. \tag{1.31.3}$$

If the bullet gets stuck in the load, then $u_1 = u_2$ and, hence,

$$u_2 = \frac{m_1 v_0}{m_1 + m_2}. \tag{1.31.4}$$

Finally, if the bullet recoils from the load, the velocity it acquires after collision, u_1, is negative and (1.31.2) can be written in the form

$$u_2 = \frac{m_1(v_0 + |u_1|)}{m_2} \tag{1.31.5}$$

A comparison of (1.31.3), (1.31.4), and (1.31.5) shows that the load acquires the highest velocity (and the greatest deflection, as a result) when the bullet recoils from it, while the lowest velocity is acquired when the bullet pierces the load.

1.32. For the sake of convenience we employ a coordinate system in which the velocity of one of the spheres prior to collision is zero. According to the energy conservation law, in the case of an absolutely elastic collision we have

$$\frac{m_1 v_0^2}{2} = \frac{m_1 u_1^2}{2} + \frac{m_2 u_2^2}{2},$$

where m_1 and m_2 are the masses of the spheres, v_0 is the velocitiy of the first sphere prior to collision, and u_1 and u_2 are the velocities of the spheres after collision. Since the masses of the spheres are the same, we can write

$$v_0^2 = u_1^2 + u_2^2.$$

The velocity vector $\mathbf{v}_0$ is the hypotenuse of a right triangle whose sides are the velocity vectors $\mathbf{u}_1$ and $\mathbf{u}_2$, and hence the angle between $\mathbf{u}_1$ and $\mathbf{u}_2$ is 90°.

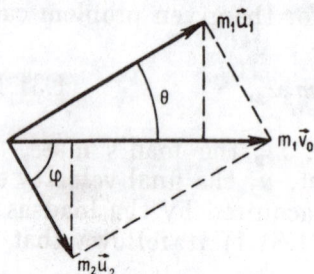

Fig. 1.33

1.33. Let u_1 and u_2 be the final velocities of the impinging sphere and the one that was at rest prior to collision, respectively, and θ is the angle between $\mathbf{u}_1$ and $\mathbf{v}_0$. The equations that express the laws of conservation of energy and momentum (for each projection) have the following form:

$$\frac{m_1 v_0^2}{2} = \frac{m_1 u_1^2}{2} + \frac{m_2 u_2^2}{2}, \tag{1.33.1}$$

$$m_1 v_0 = m_1 u_1 \cos \theta + m_2 u_2 \cos \varphi, \tag{1.33.2}$$

$$m_1 u_1 \sin \theta + m_2 u_2 \sin \varphi = 0. \tag{1.33.3}$$

If m_1, m_2, and v_0 are fixed, then u_1, u_2, θ, and φ are linked through three equations. For this reason two of the four variables can be excluded and the variable θ can be expressed in terms of the third remaining variable, say, u_1. Taking $m_1 u_1 \cos \theta$ to the left-hand side of Eq. (1.33.2), squaring the result and Eq. (1.33.3), and adding the two squares, we get

$$m_1^2 \left(v_0^2 - 2u_1u_2 \cos \theta + u_1^2 \right) = m_2^2 u_2^2.$$

Replacing u_2 with its value obtained from (1.33.1) and carrying out the necessary transformations, we arrive at a quadratic equation for u_1, namely,

$$u_1^2 - 2 \frac{m_1}{m_1+m_2} v_0 \cos \theta \times u_1 + \frac{m_1-m_2}{m_1+m_2} v_0^2 = 0, \quad (1.33.4)$$

whose solution has the form

$$u_1 = \frac{m_1}{m_1+m_2} \left(\cos \theta \pm \sqrt{\left(\frac{m_2}{m_1} \right)^2 - \sin^2 \theta} \right) v_0. \quad (1.33.5)$$

This equation shows that the maximal angle θ is determined by the condition

$$\sin \theta_m = m_2/m_1. \qquad (1.33.6)$$

For values of θ smaller than θ_m two cases are possible, since two distinct values of u_1 correspond to one value of θ. For example, for $m_1/m_2 = 3$ and $\sin \theta = 0.2$, the velocity u_1 may have two values, $0.93v_0$ and $0.53v_0$. The first collision is commonly known as soft, while the second is commonly known as hard. The extreme case of soft collision is the grazing collision (or even the case where one sphere misses the other), while the extreme case of hard collision is the head-on collision, after which the velocity of the impinging sphere becomes

$$u_1 = \frac{m_1-m_2}{m_1+m_2} v_0.$$

Condition (1.33.6) can be obtained in another manner as well. For instance, if we express $\cos \theta$ via (1.33.4), namely,

$$\cos \theta = \frac{1}{2m_1} (m_1 + m_2) \frac{u_1}{v_0} + (m_1 - m_2) \frac{v_0}{u_1},$$

and nullify the derivative of $\cos \theta$ with respect to u_1, we can find the minimal value of $\cos \theta$ or the maximal value of $\sin \theta$. The motion of the impinging sphere can also be considered using the system of coordinates linked with the center of mass of the two spheres. If in the laboratory system the coordinates of the spheres are x_1 and x_2, then the coordinate of the center of mass is

$$x = \frac{m_1 x_1 + m_2 x_2}{m_1 + m_2},$$

while the velocity of the center of mass is

$$v_c = \frac{m_1}{m_1 + m_2} v_0.$$

Correspondingly, the velocity of the impinging sphere in this system prior to collision is

$$v_0' = v_0 - v_c = v_0.$$

As a result of the collision the vector $\mathbf{v}_0'$ retains its length but turns through a certain angle depending on the distance between the center of the second sphere and

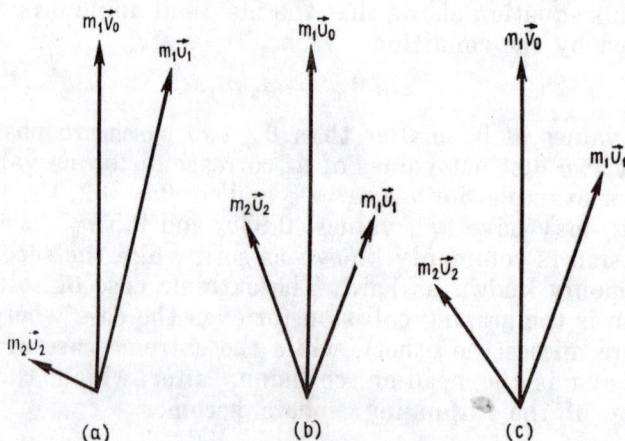

(a)　　　　(b)　　　　(c)

Fig. 1.33

the direction of flight of the impinging sphere prior to collision. The velocity $\mathbf{u}_1$ is equal to the sum of $\mathbf{v}_0'$ and $\mathbf{v}_c$. The momentum vectors of both spheres are shown in the figure for three cases: soft collision (Figure (a)) and hard collision (Figure (b)) for $m_1/m_2 = 3$ and $\sin\theta = 0.2$ and the case with $\sin\theta = m_2/m_1 = 1/3$ (Figure (c)). The velocity of the impinging sphere after collision is

$$u_1 = \frac{m_1 v_0}{m_1 + m_2} \cos\theta = 0.707 v_0.$$

The above-discussed problem is important for the theory of atomic collisions. For instance, if a potassium ion

138

impinges on a helium atom $(m_1/m_2 = 10)$, as a result of an elastic collision the ion may be deflected by an angle no greater than 5.7°.

1.34. We will consider each case in the order that it appears in the problem.

(1) The directions of the velocities of the spheres in the laboratory system are shown in the figure accompanying

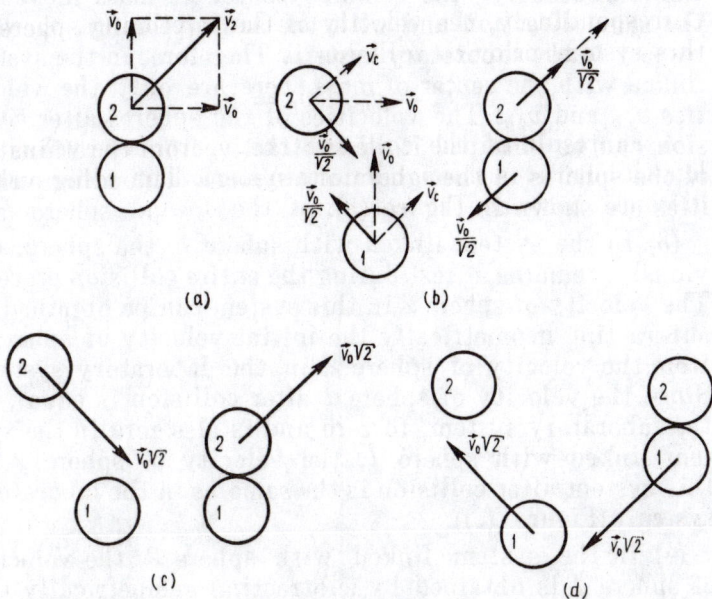

(a)   (b)

(c)   (d)

Fig. 1.34

the problem. If at the moment of collision we project the velocities of the spheres and the corresponding momenta on two axes one of which coincides with the direction of the initial velocity of sphere *1* and the other with that of the initial velocity of sphere *2*, then in the first of these two directions the spheres exchange the respective projections of the velocities, just like in a head-on elastic collision. Sphere *1* stops in the process. Since in the collision the force acts along the straight line connecting the centers of the spheres, the initial velocity of sphere *2* is conserved, with the velocity of sphere *1*, which is perpendicular to the initial velocity of sphere *2*, added to it. As a result the velocity of sphere *2* becomes equal to the geometric sum of the initial velocities of both spheres, that is, $v_0\sqrt{2}$ (Figure (a)).

(2) To determine the velocities of the spheres in the center-of-mass system, we decompose the velocity vector of each sphere into two perpendicular and equal components, v_{1a}, v_{1b} and v_{2a}, v_{2b}. The components v_{1a} and v_{2a} are equal in magnitude and point in the same direction. Obviously, the common center of mass moves in the same direction and with the same velocity, v_c, with respect to the laboratory system. Therefore, in the system linked with the center of mass there are only the velocities v_{1b} and v_{2b}. The velocities of the spheres after collision can be obtained if we subtract v_c from the velocities of the spheres in the laboratory system. The other velocities are shown in Figure (b).

(3) In the system linked with sphere *1*, the sphere, obviously, remains at rest during the entire collision process. The velocity of sphere *2* in this system can be obtained by subtracting geometrically the initial velocity of sphere *1* from the velocity of sphere *2* in the laboratory system. Since the velocity of sphere *1* after collision is equal, in the laboratory system, to zero and is also zero in the system linked with sphere *1*, the velocity of sphere *2* in this system after collision is the same as in the laboratory system (Figure (c)).

(4) In the system linked with sphere *2*, the velocity of sphere *1* is obtained by subtracting geometrically the initial velocity of sphere *2* from the velocity of sphere *1*. After collision the velocity of sphere *1* is equal, in absolute value, to the final velocity of sphere *2* in the laboratory system and points in the opposite direction (Figure (d)).

In conclusion we would like to bring the reader's attention to the fact that the angular momenta of the spheres with respect to the center of mass remain constant during the entire collision process. In collision, the center of mass is the point where the spheres touch and the angular momentum of sphere *1* is zero and remains such after collision. The angular momentum of sphere *2* is equal, prior to collision, to the product of momentum mv by the arm R. After collision the momentum of sphere *2* becomes $mv\sqrt{2}$, but the arm is now $R/\sqrt{2}$, so the product is the same and the angular momentum is conserved. Of course, since the system consisting of the spheres is isolated, the angular momentum is conserved in the entire process of motion.

1.35. After collision, sphere 2 acquires the velocity

$$u_2 = \frac{2m_1 v_1}{m_1 + m_2}. \qquad (1.35.1)$$

Sphere 3 acquires the following velocity after collision:

$$u_3 = \frac{2m_2 u_2}{m_2 + m_3}.$$

Substituting the value of u_2 from (1.35.1), we get

$$u_3 = \frac{4m_1 m_2 v_1}{(m_1 + m_2)(m_2 + m_3)}.$$

The extremal value of u_3 can be found by nullifying the derivative of u_3 with respect to m_2:

$$\frac{du_3}{dm_2} = \frac{4m_1 v_1 (m_1 m_3 - m_2^2)}{[(m_1 + m_2)(m_2 + m_3)]^2} = 0.$$

From this it follows that

$$m_2 = \sqrt{m_1 m_3}.$$

We can easily see that this value corresponds to the maximum of u_3.

Here are some particular cases.

(1) $m_1 \gg m_3$. In this case

$$u_3 \approx \frac{4m_1}{m_1 + m_2} v_1.$$

If we also assume that $m_1 \gg m_2$, then

$$u_3 \approx 4v_1.$$

If sphere 1 were to hit sphere 3 directly (without the intermediate sphere 2), the highest velocity of sphere 3 for $m_1 \gg m_3$ would be roughly $2v_1$.

In some fantastic projects of interplanetary flight it has been suggested that the spaceship be accelerated to the necessary speed through a series of collisions with intermediate objects whose masses must be calculated in the appropriate manner.

(2) $m_1 = m_3$. In this case $m_2 = m_1 = m_3$ and $u_3 = v_1$.

(3) $m_1 \ll m_3$. Assuming that $m_2 \gg m_1$, we get

$$u_3 \approx 4v_1 m_1 / m_3.$$

Here the velocity of sphere 3 is approximately double the velocity without an intermediate object, sphere 2.

1.36. The velocities of the spheres after collision are

$$u_1 = \frac{m_1 - m_2}{m_1 + m_2} v_0, \quad u_2 = \frac{2m_1}{m_1 + m_2} v_0.$$

Here are some particular cases.

(1) $u_1 < 0$ if $m_1 < m_2$. Since in this case $2m_1 < m_1 + m_2$, we have $0 < u_2 < v_0$.

(2) $u_1 = 0$ if $m_1 = m_2$. Then $u_2 = v_0$.

(3) $u_1 > 0$ if $m_1 > m_2$. Then $2m_1 > m_1 + m_2$ and $v_0 < u_2 < 2v_0$.

1.37. The equations of motion for the loads and the pulley can be written as follows:

$$m_1 w = m_1 g - T_1, \quad m_2 w = T_2 - m_2 g, \quad J\varepsilon = (T_1 - T_2)R,$$
$$(1.37.1)$$

where T_1 is the force exerted by the left end of the string on the left load, T_2 the force exerted by the right end of the string on the right load, J the moment of inertia of the pulley, w the acceleration of the loads, and ε is angular acceleration of the pulley. Dividing (1.37.1) by R, adding all the equations, and replacing ε with w/R, we arrive, after appropriate transformations, at

$$w = \frac{m_1 - m_2}{m_1 + m_2 + J/R} g. \qquad (1.37.2)$$

Equation (1.37.2) shows that in exact calculations we must allow for the moment of inertia and the radius of the pulley.

If the pulley is a homogeneous disk, then instead of J we can write $m_p R^2/2$, and Eq. (1.37.2) assumes the form

$$w = \frac{m_1 - m_2}{m_1 + m_2 + m_p/2} g.$$

We see that in this case the radius of the pulley plays no role; what is important is only the mass of the pulley.

1.38. The equations of motion for the load-shaft or the load-sheave can be written as follows:

$$mg - T = mw, \quad TR = J\varepsilon,$$

where m is the mass of each load, T the force exerted by the strings attached to the loads, R the radius of the shaft or sheave, J the moment of inertia of the shaft or sheave, and ε the angular acceleration of the shaft or sheave. Eliminating T from the equations and replacing ε with w/R and the moment of inertia of the shaft or sheave with $MR^2/2$, we arrive, after simple transformations, at

$$w = \frac{m}{m+M/2}\, g,$$

from which it follows that the accelerations with which the two loads are lowered coincide. The angular acceleration is the greater the larger the radius, which means that the shaft has a greater angular acceleration than the sheave.

1.39. Prior to switch-on, the sum of the angular momenta of all the parts of the vacuum cleaner is zero. When the motor is switched on, a torque appears in the rotor of the motor, with the same torque (in absolute value) appearing in the stator and the casing of the vacuum cleaner fixed to the stator. Due to the latter torque, the vacuum cleaner begins to turn, but this motion dies out very soon because of friction.

1.40. When the engine of the helicopter of this type is operating, two torques appear: one is applied to the main rotor and the other (equal in magnitude to the first) is applied to the fuselage of the copter. This second torque tends to turn the fuselage in the direction opposite to that of the main rotor. The vertical tail rotor creates a torque that cancels out the torque applied to the fuselage. In toy helicopters this second rotor is fixed and the helicopter rotates in flight in a direction opposite to that of the main rotor.

1.41. The rod is in rotational motion, and so its potential energy is transformed into the kinetic energy of rotation. If the mass of the rod is m and the length is l, we have

$$\frac{mgl}{2} = \frac{J\omega^2}{2}.$$

Replacing ω with v/l and J with $ml^2/3$, we get

$$v = \sqrt{3gl}.$$

1.42. To determine the trajectories that the various points of the rod describe, we introduce a coordinate system whose origin lies at B, the lower point of the rod prior to falling, whose x axis points horizontally in the direction in which point B moves during motion, and whose y axis points upward, along the rod prior to motion. Since there are no forces that act on the rod in the horizontal direction, the rod's center of mass moves downward (from C to B). As Figure (a) shows, the coordinates of the

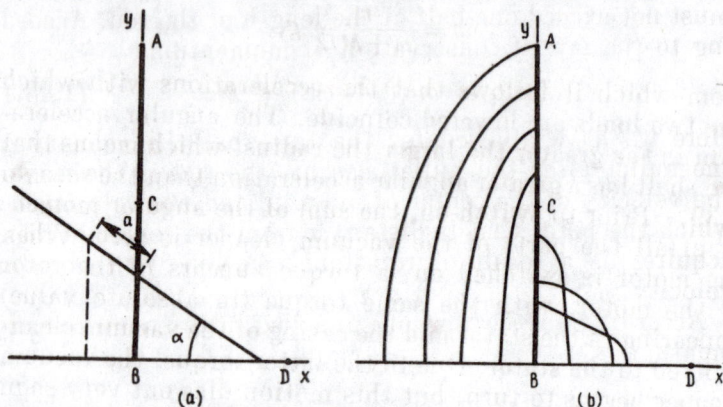

Fig. 1.42

points lying above the center of mass by a distance a are determined by the equations

$$x = -a \cos \alpha, \quad y = (R + a) \sin \alpha,$$

while the coordinates of the points lying below the center of mass by a distance a are determined by the equations

$$x = a \cos \alpha, \quad y = (R - a) \sin \alpha.$$

These equations imply that in the process of falling the rod (and that means all of its points except the center of mass) describes quarters of ellipses (Figure (b)) specified by the equations

$$\frac{x^2}{a^2} + \frac{y^2}{(R+a)^2} = 1 \quad \text{(upper points)},$$

$$\frac{x^2}{a^2} + \frac{y^2}{(R-a)^2} = 1 \quad \text{(lower points)}.$$

When the rod is falling, its motion can be considered as rotation about an instantaneous center, D. Therefore, the velocity of the upper point (A) can be determined just like in Problem 1.41, using the law of conservation of energy. The appropriate equations yield

$$v = \sqrt{6gR}.$$

1.43. The velocity imparted to point A will be directed in opposition to v_0 if the rod's linear velocity acquired as a result of rotation after the bullet has hit the rod is greater than the velocity of the center of mass of the rod. Moreover, for such a situation to occur, the distance x must not exceed one-half of the length of the rod. According to the law of conservation of momentum,

$$mv_0 = m\,(v + \omega x) + Mv. \qquad (1.43.1)$$

Here we have allowed for the fact that the velocity of the bullet after the bullet has hit the rod is the sum of the velocity of the center of mass, v, and the velocity ωx which the point that is distant x from the center of mass acquires as a result of rotational motion with angular velocity ω.

According to the law of conservation of angular momentum,

$$mv_0 x = m\,(v + \omega x)\, x + J\omega, \qquad (1.43.2)$$

where J is the moment of inertia of the rod about the center of mass, $J = MR^2/3$. Multiplying (1.43.1) by x and subtracting the product from (1.43.2), we get

$$\omega = Mvx/J = 3vx/R^2.$$

The linear velocity of rotation acquired by point A (we denote this velocity by V) is

$$V = \omega R = 3vx/R.$$

The ratio V/v is greater than unity if $x > R/3$.

1.44. According to the right-hand screw rule, the vector of the angular velocity of the gyroscope is directed to the right in the figures accompanying the problem and the answer. The revolving platform applies a torque to the frame, and the vector of this torque is directed perpendicularly to the vector of the angular velocity of the gyroscope. This torque creates an angular acceleration ε, and under this acceleration the vector of angular velocity rotates in the direction shown by the arrow in the figure accompanying the answer. As a result the giroscope's axis places itself vertically and the direction of rotation of the gyroscope coincides with the direction of rotation of the platform. If the direction of rotation of the gyroscope or the direction of rotation of the platform were to change, the gyroscope's axis would point in the opposite direction.

In all cases the axis rotates in such a manner that the vector of angular velocity places itself in the direction coinciding with that of the vector of an external torque. This property of gyroscopes is used in navigation in gyrocompasses. The "platform" that applies a torque to the gyroscope is the earth in this case.

1.45. The vector of the angular velocity of the top is directed upward along the top's axis (see the figure accompanying the answer). The force of gravity applied to the top at the top's center of mass creates a torque

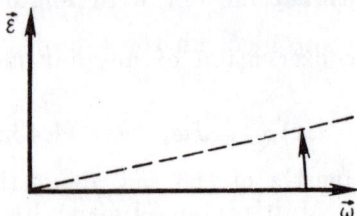

Fig. 1.44

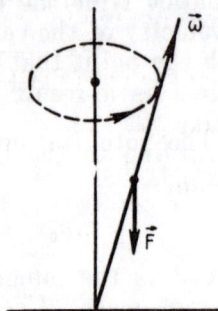

Fig. 1.45

whose vector, being perpendicular to the vector of angular velocity, is directed away from the reader. This torque does not change the magnitude of the angular velocity but creates an angular acceleration and hence changes the direction of the vector of angular velocity, just like centripetal acceleration does not change the value of the velocity but does change the direction of the velocity vector, as a result of which the body to which the centripetal acceleration is applied moves along a circle. In the case at hand the direction of the angular acceleration is such that precession occurs counterclockwise (if one views the top from above).

1.46. Since no external forces act on the shaft-sleeve system, the total angular momentum of the system remains constant:

$$J_{sh}\,\omega_0 = (J_{sh} + J_{sl})\,\omega. \qquad (1.46.1)$$

The moment of inertia of the shaft is

$$J_{sh} = \rho\,\frac{\pi d^4}{32}\,l,$$

where ρ is the density of the material of the shaft and sleeve. The moment of inertia of the sleeve is

$$J_{sl} = \rho \, \frac{\pi \, (D^4 - d^4)}{32} \, h.$$

From (1.46.1) it follows that

$$\omega_0 d^4 l = \omega \, [(D^4 - d^4) \, h + d^4 l],$$

whence

$$\omega = \frac{d^4 l}{d^4 (l - h) + D^4 h} \, \omega_0 = \frac{d^4 l}{d^4 l + (D^4 - d^4) \, h} \, \omega_0$$

$$= \frac{1}{1 + \left(\dfrac{D^4}{d^4} - 1\right) \dfrac{h}{l}} \, \omega_0.$$

1.47. The potential energy of an object on the top of a hill, mgh, transforms into the kinetic energy of translational and rotational motion:

$$mgh = \frac{mv^2}{2} + \frac{J\omega^2}{2}.$$

Replacing ω with v/R, we get

$$mgh = \frac{mv^2}{2} + \frac{Jv^2}{2R^2}. \qquad (1.47.1)$$

The moments of inertia of the disk, J_d, and the sphere, J_{sp}, are

$$J_d = \frac{mR^2}{2} \quad \text{and} \quad J_{sp} = \frac{2}{5} \, mR^2,$$

respectively, with R the radius of disk or sphere. Substituting these values into (1.47.1) and dividing by m, we get

$$gh = \frac{v^2}{2} + \frac{v^2}{4} = 0.75 v^2 \qquad (1.47.2)$$

for the disk and

$$gh = \frac{v^2}{2} + \frac{v^2}{5} = 0.7 v^2 \qquad (1.47.3)$$

for the sphere. Since the left-hand sides of these equations are the same, the final velocity of the sphere is greater, and since the motion is uniformly accelerated, the sphere will get to the horizontal section earlier than the disk. Neither the masses nor the radii of the objects rolling down the inclined planes are present in (1.47.2) and

10*

(1.47.3), with the result that the time it takes the objects to roll down is independent of these quantities.

1.48. When the spacecraft goes into a circular orbit at the perigee, it will circle the earth along a low orbit during the second half of the orbit. For this reason the spacecraft's potential energy at the new apogee will be lower than at the old one and, hence, such a maneuver requires lower kinetic energy. This means that the spacecraft must lower its velocity. Similar reasoning shows that to go into a circular orbit at the apogee, the spacecraft must increase its velocity.

1.49. The kinetic energy of a satellite is determined by the value of the orbital (or satellite) velocity. According to Newton's second law and the law of universal gravitation,

$$G \frac{Mm}{R^2} = \frac{mv^2}{R},$$

where M is the mass of the earth, m the mass of the satellite, v the velocity of the satellite, and G the gravitational constant. From this it follows that the kinetic energy

$$W_{kin} = \frac{mv^2}{2} = \frac{GMm}{2R}$$

is the smaller the higher the orbit of the satellite.

The potential energy (we take it equal to zero at infinity)

$$W_{pot} = -G \frac{Mm}{R}$$

is the greater the higher the orbit of the satellite. The same is true of the total energy:

$$W = W_{kin} + W_{pot} = -G \frac{Mm}{2R}.$$

The angular momentum also increases as we move farther away from the earth and is equal to

$$mvR = m \sqrt{GMR}.$$

1.50. Let us consider an extremely elongated orbit. In this case the distance between the foci differs little from the length of the major axis. Therefore, the force acting on a space station near the apogee can be assumed to be roughly the same for all extremely elongated orbits.

Under this force the space stations move with the same accelerations $w_c = v^2/R$, where R is the curvature radius of the trajectory, and v is the velocity at apogee. The smaller the radius of curvature, the smaller is the velocity of a space station, and the greater the elongation of the orbit, the smaller is the radius. Hence, the velocity and therefore the kinetic energy at apogee tend to zero and the space stations possess almost exclusively potential energy.

Since the total energy of a space station remains constant in flight, at all other points on the orbit it is equal to the sum of the kinetic and potential energies. The potential energy of the interaction between the earth and the station (this energy is assumed to be zero at infinity) is

$$W_{pot} = -G \frac{Mg}{a},$$

where M is the mass of the earth, m the mass of the station, G the gravitational constant, and a the distance from the center of the earth to the station (this quantity is practically equal to the length of the major axis of the orbit). When circling the earth along a circular orbit whose radius R is approximately $a/2$, the station possesses potential energy

$$W_{pot} = -2G \frac{Mm}{a}.$$

As shown in the solution to Problem 1.49, the kinetic energy of the station in this case is

$$W_{kin} = G \frac{Mm}{a},$$

while the total energy is

$$W = -G \frac{Mm}{a},$$

which means that it is the same as for an elliptical orbit.

It is convenient to determine the angular momentum of a station when the station passes through the apogee:

$$L = mva.$$

For extremely elongated orbits, a is roughly the same for all orbits, but the greater the elongation of the orbit the smaller the velocity at apogee. Hence, the angular momentum at apogee is the smaller the greater the elongation of the orbit. But since the torque of the force of attraction to the earth is zero, the angular momentum must be the same at all points of the orbit. Hence, the energy of the station in a circular orbit and that of the station in an elliptical orbit coincide, while the angular momentum is the smaller the greater the elongation of the orbit.

1.51. The fact that the spacecraft retains its orientation with respect to the earth means that all points of the spacecraft move with the same angular velocity. Suppose that the point closest to the surface of the earth moves with the orbital (satellite) velocity according to the equation

$$\omega^2 R = G\frac{M}{R^2}, \tag{1.51.1}$$

where R is the distance between this point and the center of the earth. The point of the spacecraft farthest from the earth moves with an acceleration $\omega^2 (R + D)$, where D is the distance between the two points.

If we consider the spacecraft to be a noninertial system, we can assume that on an object of mass m placed at the point farthest from the earth there acts a force of inertia

$$F_1 = -m\omega^2 (R + D).$$

At the same time, there is the force of gravity acting on this object:

$$F = G\frac{Mm}{(R+D)^2}.$$

The sum of these two forces plays the role of "weight" for the object, or numerically the reaction of the support exerted on the object:

$$F_w = m\omega^2 (R + D) - G\frac{Mm}{(R+D)^2}.$$

Bearing in mind that $D \ll R$, we can replace $(R + D)^{-2}$ with $(1 - 2D/R)/R^2$. Thus

$$F_w \approx m\left[\omega^2 R \left(1 + \frac{D}{R}\right) - G\frac{M}{R^2}\left(1 - 2\frac{D}{R}\right)\right],$$

and if we allow for (1.51.1), we get

$$F_w \approx G\frac{Mm}{R^2} - \frac{3D}{R}.$$

150

Since GMm/R^2 is equal, to a high accuracy, to the weight of the object on the surface of the earth, or mg, we get

$$F_{\mathrm{w}} = \frac{3D}{R}\, mg.$$

This expression gives the "weight" of an object in the spacecraft at the point farthest from the earth. Assuming that D is 2.1 m and bearing in mind that $R = 6300$ km, we find that the "weight" of an astronaut whose mass is 70 kg is 6.9×10^{-4} N at the point within the spacecraft farthest from the earth.

1.52. The potential energy of the comet (equal to zero at infinity) is $-GMm/r$, where m is the comet's mass, M the mass of the sun, and r the distance between the sun and the comet. As the comet approaches the sun, this energy decreases, which means that the kinetic energy increases, with

$$\frac{mv^2}{2} - G\,\frac{Mm}{r}$$

remaining zero.* The angular momentum of the comet is also conserved, since the torque produced by central forces is always zero. If we take two points, one at the aphelion of the presumable closed trajectory and the other placed at the same distance from the sun on the second branch of the parabola, then the potential energies at these points must coincide (since the distances coincide), which means that the kinetic energies at these points coincide and so do the velocities. But, as follows from the figure accompanying the problem, the angular momentum at the aphelion must be higher than on the branches of the parabola, which is impossible. At the same time, at symmetrical points both the kinetic energies and the potential energies are the same, and the same is true of the angular momenta.

The above reasoning is true for both closed orbits (ellipses and circles) and open orbits (parabolas and hyperbolas) of heavenly bodies moving in the field of a single attraction center. The fact that both the energy conservation law and the angular momentum conservation law must be satisfied makes it impossible for a central force to change the nature of a trajectory.

* It is assumed that the initial kinetic energy of the comet in far-away regions of space is negligible.

151

1.53. If D_0 is the diameter of the disk at rest, then in the system of coordinates with respect to which the disk

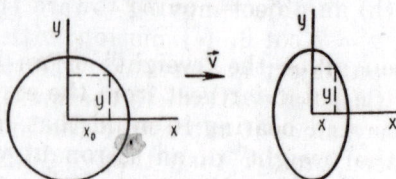

Fig. 1.53

is in motion the diameter in the direction of the velocity will be

$$D = D_0 \sqrt{1 - v^2/c^2} = D_0 \sqrt{1 - \beta^2}.$$

The same is true of the ratio of the halves of the chord passing at an altitude y from the center:

$$x = x_0 \sqrt{1 - \beta^2}.$$

Since $x_0^2 = R^2 - y^2$, we have

$$x^2 = (R^2 - y^2)(1 - \beta)^2,$$

whence

$$\frac{x^2}{R^2(1 - \beta^2)} + \frac{y^2}{R^2} = 1.$$

The moving disk appears to be an ellipse with semi-axes R and $R\sqrt{1 - \beta^2}$.

1.54. The velocity of the triangle is directed perpendicularly to the altitude, with the result that the length of the altitude is independent of the velocity. The hypotenuse is equal to twice the altitude ($l_0 = 2h$), while the length of a side of the equilateral triangle is $l = 2h \tan 30°$. Thus, for the moving triangle we have $l = l_0$ and

$$2\frac{\sqrt{3}}{3}h = 2h\sqrt{1 - \beta^2}.$$

Hence $\beta = 0.816$.

1.55. As Figure (a) accompanying this problem shows, the world line passing through the origin at an angle θ

to the x/c axis represents the motion of an object moving away from the observer (placed at the origin) with a velocity $v = c \cot \theta$. The other figures correspond to the following cases: (b) an object moving toward the observer with a velocity $v = c \cot \theta$, (c) motion with the speed of light, and (d) an object is at rest at a certain distance from the origin. Case (e) contradicts the main principles of relativity theory since it represents the motion of an object with a speed greater than that of light.

1.56. According to the theory of relativity, the kinetic energy of a moving object is given by the following formula

$$W_{rel} = m_0 c^2 \left(\frac{1}{\sqrt{1-\beta^2}} - 1 \right),$$

with $\beta = v/c$. In classical mechanics,

$$W_{cl} = \frac{m_0 v^2}{2}.$$

Thus,

$$\frac{W_{rel}}{W_{cl}} = \frac{2}{\beta^2} \left(\frac{1}{\sqrt{1-\beta^2}} - 1 \right).$$

Since $\beta = \cot \theta$, we have

$$\frac{W_{rel}}{W_{cl}} = \frac{2}{\cot^2 \theta} \left(\frac{1}{\sqrt{1-\cot^2 \theta}} - 1 \right).$$

At $\theta = 60°$,

$$W_{rel}/W_{cl} = 1.37.$$

1.57. Let us assume that at $t = 0$ by the clocks in both systems, the systems were close to each other (in the figure accompanying the problem this moment corresponds to the origin). If one of the systems sends a signal after a time interval T_0 has elapsed, the second system will receive the signal after a time interval

$$T = T_0 \sqrt{\frac{1+\beta}{1-\beta}}.$$

The angle θ corresponds to a relative velocity $\beta = \cot \theta$. Thus,

$$T = T_0 \sqrt{\frac{1+\cot \theta}{1-\cot \theta}}.$$

1.58. The time interval separating the signals received by B from A is

$$T_1 = T_0 \sqrt{\frac{1+\beta}{1-\beta}}.$$

Since system C is moving toward A, its (relative) velocity is negative and, hence, the signals it sends are received by A separated by time intervals

$$T_2 = T_0 \sqrt{\frac{1-\beta}{1+\beta}}.$$

System A will register N signals from B in the course of

$$t_1 = NT_1 = NT_0 \sqrt{\frac{1+\beta}{1-\beta}},$$

while the signals from C will be registered in the course of

$$t_2 = NT_2 = NT_0 \sqrt{\frac{1-\beta}{1+\beta}}.$$

When system A meets system C, the clock in the first system will show

$$t_B = t_1 + t_2 = NT_0 \left(\sqrt{\frac{1+\beta}{1-\beta}} + \sqrt{\frac{1-\beta}{1+\beta}} \right) = \frac{2NT_0}{\sqrt{1-\beta^2}}.$$

The clock in C will show the time that is the sum of the time during which system A sends N signals prior to meeting C and the time during which system C sends N signals prior to meeting system B. Thus,

$$t_C = 2NT_0.$$

The difference in the readings of the clocks will be

$$\Delta t = t_B - t_C = 2 \left(\frac{1}{\sqrt{1-\beta^2}} - 1 \right) NT_0.$$

The fractional variation in the duration of the signals is

$$\frac{t_B}{t_C} = \frac{1}{\sqrt{1-\beta^2}}.$$

For example, at $\beta = 0.6$ we have

$$t_B/t_C = 1.25.$$

154

2. Molecular Physics and Thermodynamics

2.1. The buoyancy, or lifting power, is the difference between the weight of the air in the volume occupied by the balloon and the weight of the gas filling the balloon. According to the ideal-gas law, the latter weight is

$$P = \frac{pVM}{RT}\, g,$$

where V is the volume of the balloon, p the pressure of the gas, and M the molecular mass of the gas. Accordingly, the lifting power is given by the formula

$$F = \frac{pVg}{RT}\, (M_{air} - M_{gas}),$$

and the buoyancy ratio is

$$\frac{F_{H_2}}{F_{He}} = \frac{M_{air} - M_{H_2}}{M_{air} - M_{He}}. \tag{2.1.1}$$

nto Eq. (2.1.1) we can substitute the relative molecular masses. The relative molecular mass of hydrogen is 2, that of helium is 4, and that of air we assume to be equal to 29. Thus,

$$\frac{F_{H_2}}{F_{He}} = \frac{29 - 2}{29 - 4} = 1.08.$$

2.2. The root-mean-square velocity of molecules is

$$v = \sqrt{3RT/M}\,.$$

Taking logs, we get

$$\log v = \frac{1}{2} \log (3R/M) + \frac{1}{2} \log T.$$

The slope of the straight line $\log v$ vs. $\log T$ must be 0.5, and the dependence of the logarithm of velocity on the logarithm of temperature is given by straight line C in the figure accompanying the problem.

2.3. Since the velocities of the molecules are different, it takes the molecules different times to fly from the slit to the outer cylinder. Because of this the cylinders rotate through angles that are different for different molecules. The greater the velocity of a molecule, the closer will its track be to the track for fixed cylinders.

2.4. The position of the tracks shown in Figure (b) accompanying the problem is possible if during the time of flight of the molecules from the slit in the inner cylinder to the wall of the outer cylinder the cylinders perform more than one-half of a full revolution (in Figure (b) this is almost one full revolution). Of course, for this to happen, the linear velocity of the outer cylinder must exceed many times the velocity of the molecules, which is practically impossible.

2.5. The number of molecules in the velocity interval from v to $v + dv$ is

$$dN = F(v)\, dv.$$

Accordingly, in Figure (a) accompanying the problem, the hatched segments represent the following quantities: segment A represents the number of molecules whose velocities do not exceed v_1, or

$$N_A = \int_0^{v_1} F(v)\, dv,$$

segment B represents the number of molecules whose velocities are not lower than v_2 and do not exceed v_3, or

$$N_B = \int_{v_2}^{v_3} F(v)\, dv,$$

and segment C represents the number of molecules whose velocities are not lower than v_4, or

$$N_C = \int_{v_4}^{\infty} F(v)\, dv.$$

In Figure (b) accompanying the problem, each hatched segment represents the ratio of the corresponding number of molecules to the total number of molecules, that is, the probability of molecules having velocities that lie within the specified velocity interval.

2.6. Since for each velocity interval from v to $v + dv$ the number of molecules is

$$dN = F(v)\, dv$$

and since $F_2(v) = 2F_1(v)$, the total number of molecules corresponding to distribution *2* is twice the number of molecules corresponding to distribution *1*.

2.7. The number of molecules in the velocity interval from v to $v + dv$ is

$$dN = F(v)\,dv.$$

Each of these molecules has an energy $mv^2/2$. All molecules in the velocity interval from v_1 to v_2 have the energy

$$W = \int_{v_1}^{v_2} \frac{mv^2}{2} F(v)\,dv.$$

To find the average energy w of such molecules, we must divide W by the number of molecules:

$$w = \frac{m}{2} \frac{\displaystyle\int_{v_1}^{v_2} v^2 F(v)\,dv}{\displaystyle\int_{v_1}^{v_2} F(v)\,dv}.$$

2.8. According to Maxwell's law, the number of molecules of a gas whose velocities lie within the interval from v to $v + dv$ is given by the formula

$$dN = N_0 4\pi \left(\frac{m}{2\pi kT}\right)^{3/2} v^2 \exp\left(-\frac{mv^2}{2kT}\right). \quad (2.8.1)$$

Since the most probable velocity is

$$v_p = \sqrt{2kT/m},$$

we can represent (2.8.1) in the form

$$dN = N_0 4\pi^{-1/2} \left(\frac{v}{v_p}\right)^2 \exp\left[-\left(\frac{v}{v_p}\right)^2\right] d\left(\frac{v}{v_p}\right).$$

The distribution function $F(v/v_p)$ then assumes the form

$$F\left(\frac{v}{v_p}\right) = N_0 4\pi^{-1/2} \left(\frac{v}{v_p}\right)^2 \exp\left[\left(-\frac{v}{v_p}\right)^2\right].$$

For $v/v_p = 1$ we have

$$F(1) = N_0 4\pi^{-1/2} e^{-1} \approx 0.83 N_0.$$

The $F(v/v_p)$-to-$F(1)$ ratio (see the figure),

$$F\left(\frac{v}{v_p}\right) \Big/ F(1) = \left(\frac{v}{v_p}\right)^2 \exp\left[1 - \left(\frac{v}{v_p}\right)^2\right],$$

157

is the same for any number of molecules of any gas at any temperature and, therefore, is a universal function.

2.9. From formula (2.9.1) it follows that

$$f(w) = \frac{1}{N_0} \frac{dN}{dw},$$

or

$$f(w) = \frac{1}{N_0} \frac{dN}{dv} \frac{dv}{dw}.$$

Since $v = (2w/m)^{1/2}$, elementary transformations yield

$$dN = N_0 \frac{2}{\sqrt{\pi}} \left(\frac{w}{kT}\right)^{1/2} \exp\left(-\frac{w}{kT}\right) d\left(\frac{w}{kT}\right).$$

This representation is convenient since the dimensionless ratio w/kT is taken as the independent variable and the

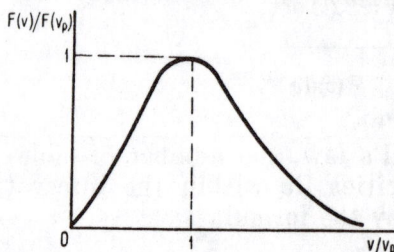

Fig. 2.8

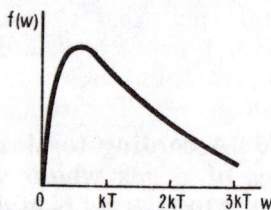

Fig. 2.9

distribution function proves to be valid not only for all gases but also for any temperature. The function $f(w)$ is shown in the figure.

2.10. The total energy of the molecules of a gas is the sum of their kinetic and potential energies. Assuming that the potential energy is zero at the initial level, for any other level we have $w_{pot} = mgh$. Since the total energy remains constant, or $w_{kin} + w_{pot} = \text{const}$, we have

$$w_{kin} + w_{pot} = w_{kino}.$$

Hence, at a given level the kinetic energy is

$$w_{kin} = w_{kino} - mgh.$$

The maximal altitude to which the molecules can rise is determined by the condition $w_{kin} = 0$, whence

$$h = w_{kino}/mg.$$

By hypothesis, $w_{k1n0} = (3/2)\,kT$. Substituting $k = R/N_A$ and $m = M/N_A$, we get

$$h = 3RT/2Mg.$$

Substituting the values of the molecular masses, we find that at $T = 300$ K the maximum altitude for nitrogen is 13.6 km, for oxygen 11.9 km, and for hydrogen 191 km. Since the kinetic energy of the molecules decreases as the altitude grows, the "temperature" of a gas decreases, too, but differently for different gases. Different gases have different "temperatures" at the same altitude above sea level. At the highest level where the molecules of a given gas can still be found, the "temperature" of the gas is 0 K.

Note, in conclusion, that by its very meaning the barometric formula, which is derived on the assumption that the temperature of the gas is constant, is equivalent to the statement that the Maxwellian velocity distribution is valid. Indeed, the barometric formula leads to Boltzmann's formula for the distribution of molecules in potential energy. The same formula can be obtained using the Maxwell formula.

2.11. To answer this question, we assume, for the sake of simplicity, that the balloon is a cylinder with its axis vertical and having a length h. If we denote by p_0 the pressure on the lower base of the balloon, then the pressure on the upper base is

$$p = p_0 \exp\,(-Mgh/RT).$$

Since $Mgh/RT \ll 1$, we can expand the exponential and retain only the first two terms:

$$p = p_0\,(1 - Mgh/RT). \qquad (2.11.1)$$

The buoyancy is given by the formula

$$F = S\,(p_0 - p),$$

where S is the base area of the cylinder. Substituting the difference $p_0 - p$ from (2.11.1), we get

$$F = MghS/RT,$$

or

$$F = p_0 MgV/RT.$$

The fraction $p_0 M/RT$ constitutes the density of air, $p_0 MV/RT$ the mass of the air that would occupy the volume of the balloon, and $p_0 MgV/RT$ the weight of this mass of air. Thus, the two explanations are equivalent.

2.12. If we take two subsequent displacements, l_1 and l_2, during time t in which these displacements took place the particle is displaced by l^*, with

$$l^{*2} = l_1^2 + l_2^2 + 2l_1 l_2 \cos \alpha$$

(see the figure accompanying the answer). Since the displacements are completely random both in length and direction, while the angle between two successive displacements is independent of the displacements, we conclude, first, that

$$\langle l_1^2 \rangle = \langle l_2^2 \rangle$$

and, second, that the third term is zero because all directions are equally probable. Thus,

$$\langle l^{*2} \rangle = 2 \langle l_1^2 \rangle = 2 \langle l_2^2 \rangle.$$

One must **bear** in mind that we have averaged the squares of the displacements and not the displacements proper.

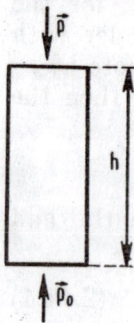

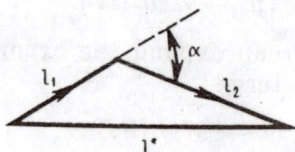

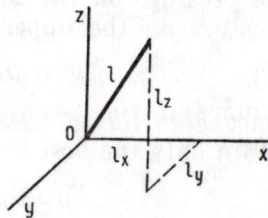

Fig. 2.11   **Fig. 2.12**   **Fig. 2.13**

However, since there is a constant relationship between the mean square and the square of the arithmetic mean for a definite distribution function, we can always replace the ratio of mean squares with the ratio of the squares of the mean,

$$\frac{(\langle l^* \rangle)^2}{2T} = \frac{(\langle l \rangle)^2}{T}$$

and, hence,

$$\langle l^* \rangle = \sqrt{2} \, \langle l \rangle.$$

This result can be applied to any interval of time, which makes it possible to establish the following relationship between the displacements of a Brownian particle and the time it takes the particle to perform these displacements:

$$\frac{(\langle l \rangle)^2}{t} = \text{const.}$$

This is the main law of Brownian motion. It is also valid for the motion of molecules in a gas.

2.13. Any concrete path of a molecule can be decomposed along three arbitrary coordinate axes of a Cartesian system, with

$$l^2 = l_x^2 + l_y^2 + l_z^2.$$

For each separate path these projections are, generally speaking, different, but since the motion is chaotic and, hence, the probabilities are the same for all three directions, these projections are equal, on the average, so that

$$\langle l_x^2 \rangle = \langle l_y^2 \rangle = \langle l_z^2 \rangle.$$

If we are interested in a projection along a definite direction, which, like all others, is arbitrary, then we can write

$$\langle l^2 \rangle = 3 \langle l_x^2 \rangle.$$

The relationship between the mean of a square and the square of a mean is the same for all directions, so that we can write

$$\langle l \rangle = \pm \sqrt{3} \langle l_x \rangle.$$

The two signs correspond to two opposite directions of motion.

2.14. If the mean free path of the molecules is λ, then the probability that on a segment dx a molecule experiences a collision will be dx/λ. Out of the N molecules that have covered the distance x without colliding, $N\,(dx/\lambda)$ molecules experience collisions over segment x. Hence, the number of molecules that have traveled without colliding will change by

$$dN = -N \frac{dx}{\lambda}.$$

If N is the total number of molecules, then the number of molecules that have traveled a distance no less than x without colliding is determined through integration:

$$\int_0^N \frac{dN}{N} = - \int_0^x \frac{dx}{\lambda},$$

or

$$\ln N = \ln N_0 - x/\lambda.$$

Since on the vertical axis we lay off base-10 logarithms, a and λ are linked in the following manner:

$$\lambda = 2.3/a.$$

Modern electronics possesses a number of methods for determining the number of particles (molecules, atoms, ions, electrons) whose path exceeds a definite distance, which makes it possible to find the mean free path.

2.15. Since the diffusion coefficient of hydrogen is higher than that of nitrogen, hydrogen will flow from

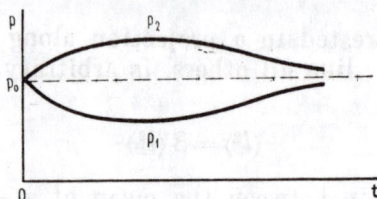

Fig. 2.15

part 1 to part 2 faster than nitrogen will flow from part 2 to part 1. For this reason, at first the pressure in part 1 drops and in part 2 it rises. But then the rate of hydrogen diffusion lowers (since the amount of hydrogen in part 2 grows and the nitrogen continues to diffuse into part 1). As a result, the pressure in part 2 begins to drop and the pressure in part 1 begins to grow. The process continues until the pressure in both parts becomes equal and the partial pressures of the two gases in each part become equal.

2.16. The diffusion coefficient of the gas is

$$D = \frac{1}{3} \lambda v.$$

In the closed vessel, the mean free path remains constant*
and the temperature dependence of the diffusion coefficient is determined only by the average velocity of the
molecules, which is proportional to the square root of
the temperature. The same relationship exists between
the temperature and the diffusion coefficient:

$$D \propto T^{1/2}.$$

In the open vessel, that is, at constant pressure, the concentration of molecules is inversely proportional to the
temperature and, hence, the mean free path is proportional
to the temperature. Therefore, for this case we have

$$D \propto T^{3/2}.$$

On the logarithmic scale the slope of a straight line is
equal to the exponent in the power function. Hence,
curve (a) (with the slope equal to 3/2) corresponds to the
open vessel and curve (b) (with the slope equal to 1/2)
corresponds to the closed vessel.

* Here we have ignored the temperature dependence of the
effective cross section (the Sutherland correction term).

2.17. The diffusion rate, which characterizes the variation of the number dN of molecules passing through the
cross-sectional area S of the vessel per unit time dt in
the direction of the concentration gradient dn/dx, is

$$\frac{dN}{dt} = -D \frac{dn}{dx} S.$$

Here $D = (1/3) \lambda v$ is the diffusion coefficient. Since the
diffusion coefficient is inversely proportional to the
pressure (because the mean free path is inversely proportional to the pressure) and the concentration gradient at
each moment is proportional to the pressure, the number
of molecules diffusing in this or that direction is pressure
independent. This conclusion holds, of course, only if the
mean free path of the molecules is many times smaller
than the linear dimensions of the vessel. Note that since
the initial number of molecules of each gas is proportional
to the pressure, the evening out of the concentrations
occurs the faster the lower the pressure of the gas.

2.18. The average kinetic energy of translational motion
of molecules is $(3/2)\,kT$. The average energy of the molecules moving toward a wall of the vessel is $2kT$. This is
explained by the fact that the flux of molecules with

a certain velocity is proportional to $n_v v$, where n_v is the concentration of the molecules having this velocity. Therefore, the higher the velocity, the greater the number of molecules moving in a given direction. Hence, in the velocity distribution of the molecules remaining in the vessel there appears a deficit of fast molecules, which leads to a decrease in the average energy of the molecules and a distortion in the distribution function. On the other hand, the average energy of the molecules leaving the vessel for the vacuum becomes higher than it was in the vessel. If the pressure of the gas is not low but the orifice is so small that no collisions occur in it, the average energy inside the vessel still decreases, if only this decrease is not compensated for by heat supplied to the walls of the vessel. Under these conditions, the Maxwellian velocity distribution is restored via the collisions of molecules in the vessel, but now this distribution corresponds to a lower temperature. The restoration of the distribution function occurs partially because molecules collide with the walls of the vessel.

2.19. The heat flux is determined by the relationship

$$\frac{dQ}{dt} = -\lambda \frac{dT}{dx}.$$

For the thermal conductivity of an ideal gas we have the following formula:

$$\lambda \propto v, \quad \text{or} \quad \lambda \propto T^{1/2}.$$

For the flux to be steady-state (time independent), the following formula must hold true:

$$\lambda \frac{dT}{dx} = \text{const.}$$

Hence,

$$T^{1/2} \frac{dT}{dx} = \text{const.}$$

We see that the higher the temperature the lower is the gradient. The gradient must increase from the hot plate to the cold plate. The position of the plates can be explained by the necessity of reducing convection to a minimum.

2.20. Under the specified conditions, we cannot apply the concept of temperature to the residual gas between the walls of the Dewar vessel. The mean free path of the

molecules of the gas is about 100 m, so that while moving
between the walls the molecules practically never collide
with each other and no thermodynamic equilibrium,
which could be characterized by a temperature, can establish itself between the walls.

2.21. Within a broad pressure range the thermal conductivity coefficient is independent of the gas pressure.
A dependence (i.e. a drop in thermal conductivity as the
pressure lowers) becomes noticeable if the mean free path
of molecules becomes comparable to the distance between
the walls between which the heat transfer occurs. The
greater this distance, the greater the mean free path (and
the lower the pressure) at which the thermal conductivity
coefficient begins to change. Therefore, curve *1* corresponds to the greater distance (see the figure accompanying
the problem).

2.22. Section *1-2* in Figure (a) accompanying the problem corresponds to isobaric heating, section *2-3* to

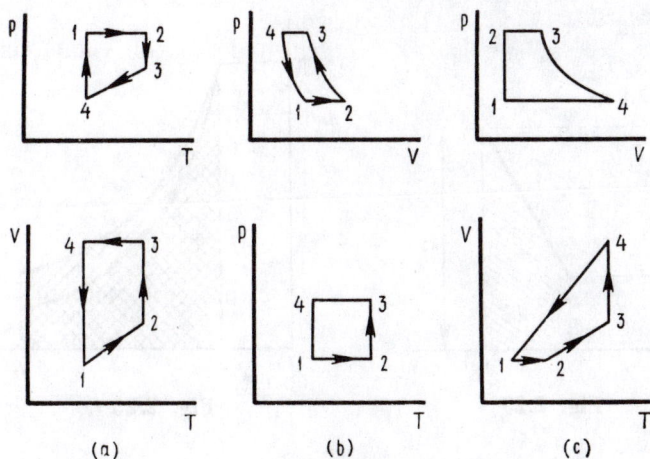

Fig. 2.22

isothermal expansion, section *3-4* to isochoric cooling,
and section *4-1* to isothermal compression. In the *pT*-
and *VT*-coordinates this process is depicted in Figure (a)
accompanying the answer. The processes depicted in Figure (b) accompanying the problem proceed in the following order: *1-2* is isobaric heating, *2-3* isothermal compression, *3-4* isobaric cooling, and *4-1* isothermal expan-

sion. In the pV- and pT-coordinates this cycle is depicted in Figure (b) accompanying the answer. The cycle depicted in Figure (c) accompanying the problem consists of isochoric heating *1-2*, isobaric heating *2-3*, isothermal expansion *3-4*, and isobaric cooling *4-1*. In the pV- and VT-coordinates this cycle is depicted in Figure (c) accompanying the answer.

2.23. When the piston moves upward by Δx, the spring is compressed by Δh. Suppose $F = -k\Delta h$ is the elastic force produced in the spring by this compression. This force contributes to the force acting on the piston and, hence, increases the pressure of the gas in the cylinder by

$$\Delta p = \frac{|F|}{S} = \frac{k\,\Delta h}{S} = \frac{k\,\Delta V}{S^2},$$

where S is the surface area of the piston. Thus, the increase of the gas volume caused by heating is accompanied by a proportional increase in the pressure. On the dia-

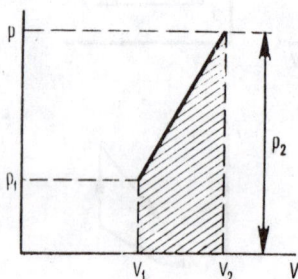

Fig. 2.23

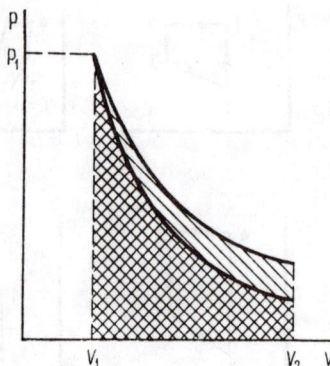

Fig. 2.25 ·

gram this is depicted by a straight line with a positive slope whose value depends on the surface area of the piston and the elastic properties of the spring. The work is measured by the hatched area in the figure accompanying the answer and is

$$A = (p_1 + p_2)(V_2 - V_1)/2.$$

2.24. The adiabatic p-V relation is of the form

$$pV^\gamma = \text{const},$$

where the exponent γ is the ratio of the specific heat capacity of the gas at constant pressure to the specific heat capacity of the gas at constant volume:

$$\gamma = c_p/c_V.$$

This ratio can be expressed in terms of the number of degrees of freedom, i. A helium molecule has three degrees of freedom and that of carbon dioxide has six. Therefore, for helium we have $\gamma = 5/3 = 1.67$ and for carbon dioxide we have $\gamma = 8/6 = 1.33$. The greater the exponent, the steeper is the curve. The upper curve (see the figure accompanying the problem) corresponds to carbon dioxide and the lower curve corresponds to helium.

2.25. An adiabatic curve is steeper than an isotherm (see the figure accompanying the answer), with the final pressure being lower in the adiabatic process than that in the isothermal process. This means that the area lying below the appropriate curve (this area characterizes the work) is smaller for the adiabatic process than for the isothermal.

2.26. The first law of thermodynamics for an isothermal process can be written in the form $Q = A$. Hence, the straight line corresponding to this process must be inclined at an angle of 45° to the horizontal axis (curve 3 in the figure accompanying the problem). For an isobaric process we have

$$Q = \Delta U + A.$$

Since the work for one mole of the gas done in an isobaric process is

$$A = R\Delta T$$

and the amount of the absorbed heat is

$$Q = C_p \, \Delta t = \frac{i+2}{2} \, R\Delta T,$$

with i the number of degrees of freedom, we have

$$\frac{A}{Q} = \frac{2}{i+2}.$$

The slope of the straight line representing the A vs. Q dependence must equal 2/5 for a monatomic gas, 2/7 for a diatomic gas, and 2/8 for a multiatomic gas. Straight line 1 corresponds to a multiatomic gas and straight line 2

to a monatomic gas. Work is not performed in an iso-choric process, and this coincides with the horizontal axis, while heat is not absorbed in an adiabatic process, and this coincides with the vertical axis.

2.27. Processes depicted by straight lines coinciding with the coordinate axes are quite obvious. The horizontal axis ($\Delta T = 0$) represents an isothermal process and the vertical axis ($Q = 0$) represents an adiabatic process. The molar heat capacity of a monatomic gas involved in an isochoric process is

$$C_V = (3/2)\,R,$$

and that of a diatomic gas is

$$C_V = (5/2)\,R.$$

The molar heat capacity of a monatomic gas involved in an isobaric process is

$$C_p = (5/2)\,R,$$

and that of a diatomic gas is

$$C_p = (7/2)\,R.$$

The heat capacity C_V of a diatomic gas coincides with the heat capacity C_p of a monatomic gas. For this reason there are three straight lines in the figure accompanying the problem instead of four, with straight line *2* corresponding to C_p of a monatomic gas and C_V of a diatomic gas. Straight line *3* corresponds to an isobaric process involving a diatomic gas and straight line *1* corresponds to an isochoric process involving a monatomic gas.

2.28. For the sake of brevity we denote $(m/M)\,R$ by a. Then

$$pV = aT.$$

For both gases the work performed in an isobaric process is

$$A = p\,(V_2 - V_1) = a\Delta T,$$

while that performed in an adiabatic process is

$$A = \frac{p_1 A_1 - p_2 V_2}{\gamma - 1} = a\,\frac{T_1 - T_2}{\gamma - 1} = a\,\frac{|\,\Delta T\,|}{\gamma - 1}.$$

Substituting the values of γ for nitrogen (7/5) and argon (5/3), we get

$$A = 2.5 \mid \Delta T \mid \quad \text{(for nitrogen)},$$

$$A = 1.5 \mid \Delta T \mid \quad \text{(for argon)}.$$

Selecting the scales on the coordinate axes of the figure accompanying the problem in such a manner that $a = 1$, we find the slopes of the straight lines to be 2.5 and 1.5 for the adiabatic processes and 1 for the isobaric process. The straight line *1* depicts the adiabatic process involving nitrogen, the straight line *2* depicts the adiabatic process involving argon, and the straight line *3* depicts the isobaric process for both gases. The vertical axis ($\mid \Delta T \mid = 0$) depicts an isotherm and the horizontal axis ($A = 0$) an isochor.

2.29. The classical theory of heat capacity does not allow for the quantum nature of periodic motion (vibrational and rotational). According to quantum theory, the angular momentum of a rotating object may assume only values specified by the condition

$$J\omega = \hbar \sqrt{j(j+1)}, \qquad (2.29.1)$$

where $\hbar$ is the Dirac-Planck constant (the Planck constant h divided by 2π), J is the moment of inertia of the object, and j is the so-called rotational quantum number, which can take on any integral values starting from zero. Equation (2.29.1) enables finding the possible values of the rotational kinetic energy:

$$W = \frac{J\omega^2}{2} = \frac{\hbar^2}{2J} j(j+1).$$

The minimal nonzero value is

$$W_{\mathrm{m}} = \hbar^2/J.$$

A molecule acquires and exchanges rotational energy through collisions with other molecules. Thus, the question of whether a molecule can have rotational energy in addition to translational is solved by comparing the value W_{m} of the minimal nonzero rotational energy with a quantity of the order of kT at room temperature. The separation of atoms in a hydrogen molecule is 0.74 nm and the mass of each atom is 1.67×10^{-27} kg, so that the moment of inertia of a hydrogen molecule is $4.6 \times$

10^{-48} kg·m². Bearing in mind that $\hbar = 1.05 \times 10^{-34}$ J·s, we get

$$W_m = 2.4 \times 10^{-21} \text{ J.}$$

At room temperature $(T \approx 300 \text{ K})$,

$$kT = 4.1 \times 10^{-21} \text{ J.}$$

The fact that kT somewhat exceeds W_m makes the occurrence of rotational motion in a molecule quite probable. Hence, the rotational degrees of freedom will contribute to the heat capacity of hydrogen. At temperatures of the order of 40 K the probability of rotational motion is practically nil; it is said that the rotational degrees of freedom "freeze out" and only the translational degrees of freedom remain, which is reflected in the value of the heat capacity. The diatomic gas that is closest to hydrogen in the Periodic Table is nitrogen, and the mass of a nitrogen atom is fourteen times the mass of a hydrogen atom. The separation of the atoms in a nitrogen atom is 0.11 nm. Accordingly, the moment of inertia of a nitrogen molecule is thirty one times that of a hydrogen molecule, so that down to very low temperatures the value of kT is considerably higher than W_m and there is practically no "freezing out" of rotational degrees of freedom. At the same time, for monatomic gases, whose moment of inertia is several orders of magnitude lower than that of hydrogen, the minimal energy of rotational motion is so high that even at very high temperatures only the translational degrees of freedom manifest themselves and the heat capacity follows the predictions of classical theory quite accurately.

2.30. According to classical theory of heat capacity of ideal gases, the value of heat capacity for each given process (say, an isochoric process) must not depend on the temperature of the gas. This theory does not allow for the quantum nature of periodic processes, namely, rotational and vibrational motion. In classical theory, the probability of rotational motion of diatomic and multiatomic molecules is assumed to be independent of the temperature of the gas and the same (per each degree of freedom) as that of translational motion. Quantum theory requires allowing for the different probabilities of periodic processes, with the probability growing with temperature. Calculations have shown that for many diatomic gases at low temperatures the vibrational degrees of freedom can be ignored, but the role of these degrees of freedom grows with temperature. For sufficiently high temperature the bonds between the atoms may break and dissociation occurs. This requires large energy expenditure. In some respects this process resembles phase transitions (melting and boiling, for instance), when supplying heat does not lead to a rise in temperature.

2.31. The compressibility is defined by the following formula:

$$\beta = -\frac{1}{V}\frac{dV}{dp}.$$

In an isothermal process,

$$d\,(pV) = p\,dV + V\,dp$$

and, hence, $\beta = 1/p$. In an adiabatic process,

$$\gamma p V^{\gamma-1}dV + V^{\gamma}dp = 0$$

and, hence, $\beta = 1/\gamma p$.

In all cases the dependence of the compressibility on pressure is depicted by hyperbolas that differ only in

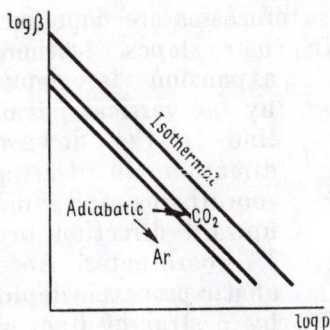

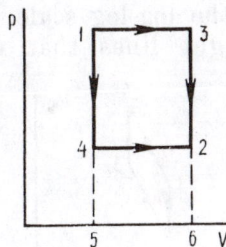

Fig. 2.31   Fig. 2.32

a numerical factor. On the log-log scale the pressure dependence of the compressibility is depicted by straight lines (in the figure accompanying the answer the straight lines correspond to adiabatic processes involving argon and carbon dioxide and to the isothermal process).

2.32. In the figure accompanying the answer the processes are depicted by broken lines *1-3-2* and *1-4-2*. In the first case, the work is measured by the area bounded by the broken line 5-4-2-6, while in the second it is measured by the area bounded by the broken line 5-1-3-6 and exceeds the first area by the area of 4-1-3-2-4. Since in both cases the initial states (1) and the final states (2) are the same, the increment of internal energy is the same, too, but the process 1-3-2 requires additional heat for the system to perform greater work. Since entropy is a function of state, the change of entropy in both cases is the same.

2.33. The three quantities characterizing the state of an ideal gas, p, V, and T, are linked through power relationships for all processes involving an ideal gas:

$$V/T = \text{const} \quad \text{(isobaric process)},$$
$$p/T = \text{const} \quad \text{(isochoric process)},$$
$$pV = \text{const} \quad \text{(isothermal process)},$$

$$\left.\begin{array}{l} pV^{\gamma} = \text{const} \\ VT^{1/(\gamma-1)} = \text{const} \\ p/T^{\gamma/(\gamma-1)} = \text{const} \end{array}\right\} \quad \text{(adiabatic process)}.$$

On the log-log scale all these processes are depicted by straight lines that differ in their slopes. Isothermal expansion is depicted by a vertical straight line in the downward direction in the pT-coordinates and in the upward direction in the VT-coordinates. An adiabatic process is depicted by a straight line with a slope $\gamma/(\gamma - 1)$ in the pV-coordinates and by a straight line with a negative slope $-(\gamma - 1)^{-1}$ in the VT-coordinates. A collection of such straight segments can be used to depict the Carnot cycle in the pT-coordinates (Figure (a)) and in the VT-coordinates (Figure (b)).

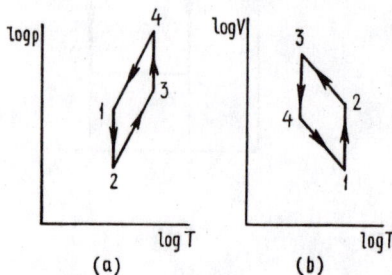

Fig. 2.33

2.34. The increment of the entropy in a process is given by the formula

$$\Delta S = \Delta Q/T.$$

172

The straight line *0-1* in the figure accompanying the problem corresponds to an isothermal process, since it is parallel to the vertical axis ($T = \text{const}$). The straight line *0-4* depicts a process in which the entropy does not change, that is, a process in which no heat is supplied to or removed from the system, or an adiabatic process. Out of curves *0-2* and *0-3* the former corresponds to a higher entropy increment. The process represented by this curve will require a larger amount of heat for bringing the system to a given temperature than the process represented by curve *0-3* will require for bringing the system to the same temperature. Of two processes, the isochoric and the isobaric, the latter requires more heat to perform work on the system. Thus, curve *0-2* corresponds to an isobaric process and curve *0-3*, to an isochoric.

2.35. The first process in the Carnot cycle is isothermal expansion. In the process the gas absorbs heat and its entropy increases. On the diagram this process is shown by the straight line *1-2*. This is followed by adiabatic expansion, which is accompanied by a drop in temperature. Since in an adiabatic process the gas is thermally isolated, the entropy cannot change, which is represented by the straight line *2-3*. At the temperature achieved at the

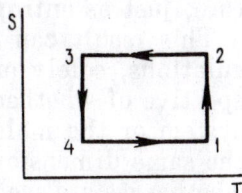

Fig. 2.35

end of this process the third process begins, namely, isothermal compression, in which the gas gives off heat and its energy decreases (the straight line *3-4*). The final process is adiabatic compression, which returns the gas to the initial state. The entropy does not change in this last process, just as in adiabatic expansion. The process is depicted by the straight line *4-1*.

2.36. If m is the mass of each object and c is the specific heat capacity, then the total entropy increment is

$$\Delta S = \int_{T_1}^{T_2} \frac{dQ}{T} + \int_{T_2}^{T_1} \frac{dQ}{T} = cm \left(\ln \frac{T}{T_1} + \ln \frac{T}{T_2} \right).$$

Replacing T with $(T_1 + T_2)/2$, we can write

$$\Delta S = cm \ln \frac{(T_1 + T_2)^2}{4T_1 T_2},$$

173

or

$$\Delta S = \Delta S_1 + \Delta S_2 = cm \ln \left[\frac{(T_1 - T_2)^2}{4T_1 T_2} + 1 \right].$$

The expression in square brackets is greater than unity, with the result that

$$\Delta S > 0.$$

2.37. The entropy increment in the process is

$$\Delta S = \Delta Q / T = c \Delta T / T. \qquad (2.37.1)$$

According to the figure accompanying the problem,

$$\mathrm{d}S = a \, \mathrm{d}T. \qquad (2.37.2)$$

The straight line in the figure passes through the origin since by the Nernst heat theorem the entropy at $T = 0$ is zero. Combining these two equations, we get

$$a \, \mathrm{d}T = c \, \mathrm{d}T / T \quad \text{and} \quad c = aT.$$

The heat capacity changes in proportion to the temperature, just as entropy does.

This result can be obtained without carrying out calculations, solely on dimensional grounds. Entropy (irrespective of whether we are speaking of the entropy of the system or the molar entropy or the specific entropy) has the same dimensionality as heat capacity (irrespective of whether we are speaking of the heat capacity of the system or the molar heat capacity or the specific heat capacity). For this reason the dependence of heat capacity on temperature must be the same, to within a constant factor, as the dependence of entropy on temperature. In the case at hand the constant factors coincide, too.

2.38. The entropy increments on different segments are

$$\Delta S_{1\text{-}2} = c_p \int_{T_1}^{T_2} \frac{\mathrm{d}T}{T} = c_p \ln \frac{T_2}{T_1} = c_p \ln \frac{V_2}{V_1},$$

$$\Delta S_{1\text{-}3} = c_V \int_{T_1}^{T_3} \frac{\mathrm{d}T}{T} = c_V \ln \frac{T_3}{T_1} = -c_V \ln \frac{p_1}{p_3} = -\ln \frac{p_2}{p_4},$$

$$\Delta S_{3\text{-}4} = c_p \int_{T_3}^{T_4} \frac{\mathrm{d}T}{T} = c_p \ln \frac{T_4}{T_3} = c_p \ln \frac{V_4}{V_3} = c_p \ln \frac{V_2}{V_1},$$

$$\Delta S_{4\text{-}2} = c_V \int_{T_4}^{T_2} \frac{\mathrm{d}T}{T} = c_V \ln \frac{T_2}{T_4} = c_V \ln \frac{p_2}{p_4}.$$

If we add all four quantities, we get

$$\Delta S_{1\text{-}3\text{-}4\text{-}2} = \Delta S_{1\text{-}3} + \Delta S_{3\text{-}4} + \Delta S_{4\text{-}2} = c_p \ln \frac{V_2}{V_1} = \Delta S_{1\text{-}2},$$

which is what we set out to prove.

2.39. For the sake of making the calculations shorter, let us select the mass in such a manner that in the appropriate system of units $(m/M)\, R = 1$. In this case the temperature of the heater, which is the highest temperature in the cycle (point c) is

$$T_h = p_2 V_2.$$

The temperature of the cooler (the coldest point in the cycle is point a) is

$$T_c = p_1 V_1.$$

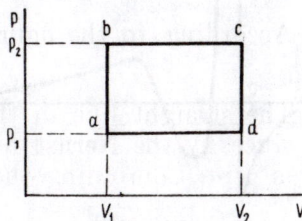

Fig. 2.39

The temperatures at points b and d are $p_2 V_1$ and $p_1 V_2$, respectively. The entropy increment for the heater in the a-b process is

$$\Delta S_{a\text{-}b} = - \frac{c_V V_1 (p_2 - p_1)}{p_2 V_2},$$

and in the b-c process it is

$$\Delta S_{b\text{-}c} = - \frac{c_p p_2 (V_2 - V_1)}{p_2 V_2}.$$

The entropy increment for the cooler in the c-d process is

$$\Delta S_{c\text{-}d} = + \frac{c_V V_2 (p_2 - p_1)}{p_1 V_1},$$

and in the d-a process it is

$$\Delta S_{d\text{-}a} = + \frac{c_p p_1 (V_2 - V_1)}{p_1 V_1}.$$

Adding all these entropy increments and carrying out the necessary transformations, we get

$$\Delta S = c_V (p_2 - p_1) \left[\frac{V_2}{p_1 V_1} - \frac{V_1}{p_2 V_2} \right]$$
$$+ c_p (V_2 - V_1) \left(\frac{1}{V_1} - \frac{1}{V_2} \right).$$

All the differences in the brackets are positive, and hence

$$\Delta S > 0.$$

The entropy increment for a gas performing a cycle and returning as a result to the initial state is equal to zero.

2.40. If we solve the van der Waals equation for p, we get

$$p = \frac{m}{M} \frac{RT}{(V - b)} - \frac{a}{V^2}.$$

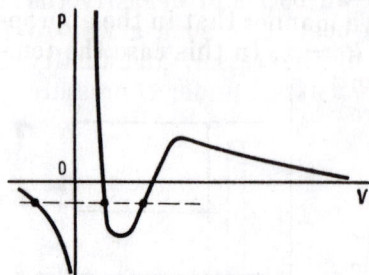

This equation is of a hyperbolic nature, and because of this it must have a branch in the third quadrant, which contains, at a sufficiently low temperature and a negative pressure, the third root. Since this root corresponds to a negative volume, it has no physical meaning and is usually not depicted on diagrams. Note that Boyle's law also contains an "extra" root. It also lies in the third quadrant and for this reason has no physical meaning and is usually not depicted on diagrams.

Fig. 2.40

2.41. The van der Waals equation presupposes complete homogeneity of the substance (vapor or liquid), that is, the same density in all (however small) volumes. In real media, however, there are fluctuations. Suppose we are considering state 2 on the curve (see the figure accompanying the problem). The parameters of this state (or point) determine the average values of the concentration and energy of the molecules. In small volumes the values of the concentration are somewhat larger or smaller than the average value because of the randomness of molecular motion. The same is true of the energy of molecules in small volumes. In accordance with the isotherm, in volumes of higher density the pressure is somewhat lower than the average, while in volumes of lower density the pressure is somewhat higher. Therefore, in the former the density continues to rise and in the latter, to drop. As a result the entire substance separates into two phases with a higher and a lower density, and the pressure in both is the same. The greater density is that of the liquid and the lower is that of the saturated vapor of this liquid.

2.42. Section *2-3* corresponds to supersaturated vapor. For this state to realize itself, there must be no dust, ions, or aerosols in the space where this state occurs for the vapor to condense on and form drops of liquid. Section *6-5* corresponds to the so-called superheated liquid. This state can be arrived at if we boil and degasify the liquid prior to heating it, then heat it in such a way so that it fills the entire volume of the vessel, and finally cool it again. The liquid will find itself under a pressure that is lower than that of the saturated vapor. Especially interesting is the state of a liquid corresponding to the section of the isotherm lying below the horizontal axis (see the figure accompanying Problem 2.40). This state corresponds to uniform stretching of the liquid. The state can be achieved by repeating, say, Torricelli's experiment in modern vacuum conditions. Before filling the tube with mercury, all gases must be evacuated from the tube via prolonged heating and the mercury must be pumped into the tube under a vacuum. In this case there is no Torricellian vacuum above the mercury when we turn the tube over, the mercury sticks to the inner surface of the tube thanks to molecular adhesion, and the part of it lying above the level corresponding to atmospheric pressure will be under negative pressure (see the figure accompanying the answer). Thus, it is possible to obtain negative pressure (uniform stretching) of the order of three atmospheres.

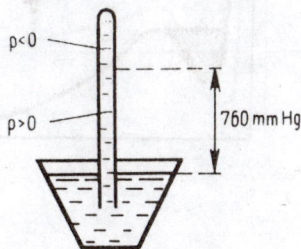

Fig. 2.42

2.43. We use the reductio ad absurdum proof. In Figures (a) and (b) accompanying the answer we have two variants that differ from the variant shown in the figure accompanying the problem. In each of these variants the arrows show a cyclic isothermal process. As a result of each of these processes, useful work is done (the amount of this work is equal to the hatched area) with an efficiency of 100% thanks to complete utilization of the heat received from the heater, which of course contradicts the second law of thermodynamics and, hence, is impossible. If we assume that the hatched areas in the figure accompanying the problem are the same, the works done along the paths *2-4-6* and *2-3-4-5-6* are equal. But doesn't this contradict the second law of thermodynamics, that is, can a cyclic process along the path *2-4-3-2* be per-

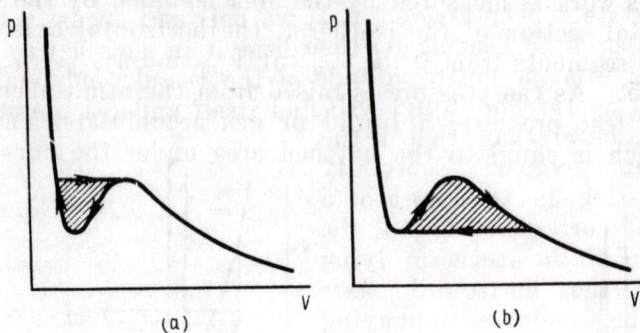

Fig. 2.43

formed? One must bear in mind that while points *2* and *6* correspond to a single (i.e. the same for both curves) one-phase state, point *4* corresponds to two different states, a one-phase state on the theoretical curve and a two-phase state on the experimental curve. The entropies of these two states are different, and so are the internal energies of these states, energies related to the interaction between the molecules.

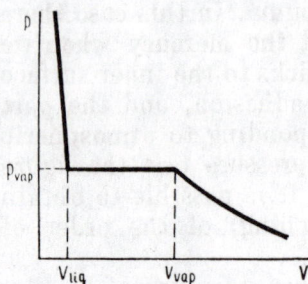

Fig. 2.44

2.44. Suppose that under the piston there is a liquid and its saturated vapor, whose pressure is counterbalanced by the external pressure. If heat is supplied to the liquid isothermally, the liquid evaporates and the piston rises. The work done by the vapor when the vapor increases its volume by ΔV is given by the formula

$$A = p_{\mathrm{vap}}\Delta V,$$

where p_{vap} is the pressure of the saturated vapor. If at the beginning there is only liquid whose volume is V_{liq}

and at the end only vapor whose volume is $\dot{V}_{\text{vap}}$, the entire work done during evaporation is

$$A = p_{\text{vap}} (V_{\text{vap}} - V_{\text{liq}}).$$

This work is measured by the area bounded by the horizontal section of the isotherm, the horizontal axis, and the segments from 0 to p_{vap} at V_{liq} and V_{vap}.

2.45. As the pressure is raised from the atmospheric to the test pressure, a liquid or gas accumulates energy, which is equal to the hatched area under the curve. If

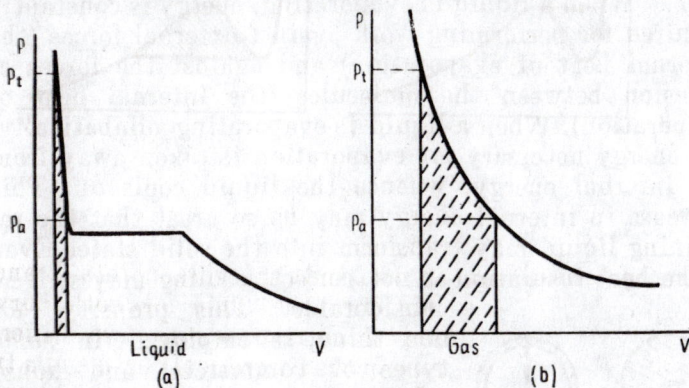

Fig. 2.45

the cylinder or pipe fails, only a small fraction of the energy is liberated by the liquid (Figure (a)) because of the small compressibilities of liquids, and the pressure falls to the atmospheric practically immediately. In the case of a gas the accumulated energy may be extremely high (Figure (b)) and the consequences of its liberation may be catastrophic.

2.46. When a liquid is heated, its density drops, so that the volume it occupies may increase notwithstanding evaporation. The decrease in the density of the liquid and the simultaneous increase in the density of the vapor lead to a drop in surface tension. As a result the meniscus becomes flatter and at the critical point disappears completely. Of special interest is the phenomenon of critical opalescence, discovered by T. Andrews in 1869, which consists in the medium becoming suddenly "cloudy" at the critical state. This phenomenon serves as a vivid illustration of fluctuation effects. Extremely small fluctuations in the density of the medium, fluctuations that are due to the random movements of molecules, lead to a situation in which the density in some microscopic volumes becomes, at certain moments, somewhat higher than the one corresponding to the critical point, and these volumes transform into the liquid, while the neighboring volumes remain being a gas (the ones with the lower density). In subsequent moments this situation may change. In this sense the entire volume filled with the fluid consists of constantly changing liquid-gas interfaces on which the light is scattered.

2.47. When a liquid is evaporating, energy is constantly required for performing work against external forces (the external heat of evaporation) and against the forces of cohesion between the molecules (the internal heat of evaporation). When a liquid is evaporating adiabatically, the energy necessary for evaporation is taken away from the internal energy, whence the liquid cools off. This decrease in internal energy may be so great that the remaining liquid may transform into the solid state. Even if the heat insulation is not perfect, cooling may still be considerable. This property, for one thing, is employed in some types of commercial and house refrigerators.

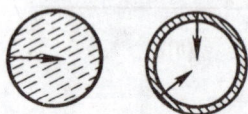

Fig. 2.48

2.48. A drop has only one spherical surface while a bubble has two, the inner and the outer, whose curvatures are almost the same in magnitude but opposite in sign. For this reason the two surfaces of a bubble create excess pressure directed toward the center of the bubble. Thus, the excess inner pressure in a bubble is approximately twice as large as in a drop (of the same radius).

2.49. The excess pressure inside a bubble is determined by the formula

$$\Delta p = 4\sigma/r,$$

where r is the radius of the bubble, and σ is the surface tension. Because of this the pressure inside the smaller bubble is greater and the bubble contracts, while the larger bubble grows. Equilibrium is attained when the film of the smaller bubble forms a surface near the outlet of the pipe with a curvature radius that coincides with the one of the larger bubble.

2.50. The vapor pressure above the convex surface of a liquid is higher than that above the flat surface, with the corresponding difference being the greater the smaller the curvature radius of the surface.* Hence, for the smaller drop (Figure (c)) the vapor is unsaturated, while for the greater drop (Figure (a)) the vapor is supersaturated. Drop (a) evaporates, while drop (c) grows. The equilibrium of drop (b) is unstable, since if the size somewhat decreases, the drop begins to evaporate, while if the size increases, the drop grows.

* The excess pressure is determined via the Thomson formula

$$\Delta p = \frac{p_{vap} M \sigma}{RT \rho r},$$

where p_{vap} is the vapor pressure above the surface, M the molecular mass (weight), R the universal gas constant, T the temperature, σ the surface tension of the liquid, ρ the density of the liquid, and r the curvature radius of the surface.

2.51. The curvature of the surface of a liquid creates an excess pressure (known as Laplace pressure) directed toward the center of curvature. This pressure is the higher the smaller the radius of curvature of the surface. In the case of water, the excess pressure (negative) tends to stretch the drop, while in the case of mercury it tends to compress the drop. For this reason, the plates with the drop of water between them are under forces that bring them together ("attractive forces"), while the plates with mercury between them tend to move apart ("repulsive forces").

2.52. The excess Laplace pressure, caused by the curvature of the liquid surface, is directed toward the center of curvature of the surface and is inversely proportional to the radius of curvature. For this reason, the drop of water is under a negative pressure (that in absolute value is greater than the pressure acting on the mercury drop) in the narrow part of the pipe and this pressure is directed toward the tapered end, with the result that the drop tends to move toward the tapered end. In the case of mercury, the pressure is directed in opposition, that is, toward the wide end of the pipe, and it is in this direction that the drop tends to move.

2.53. Surface tension (the surface tension coefficient) is defined as the ratio of the free energy of the surface layer of the liquid to the area of this surface. The free energy here is understood to be the energy that can be converted into work. This energy is determined by the interaction of the molecules of the surface layer with the other molecules, where the interaction with the molecules of the vapor above the surface is usually ignored. As the temperature is increased, the interaction of the molecules of the surface layer with the molecules in the bulk of the liquid weakens and that of the surface layer molecules with the vapor molecules grows. At the critical temperature both interactions become equal, the interface between liquid and vapor disappears, and so does surface tension. Thus, it is curve 2 that reflects the correct temperature dependence of the surface tension coefficient.

Fig. 2.54

2.54. If we assume that the water wets the wall of the tube in an ideal manner, then, if the tube is sufficiently high $(h > h_0)$, and the diameter of the tube is small, the radius of the meniscus is equal to that of the tube. If $h < h_0$, the water will rise in the tube and reach the upper end. After this the curvature of the meniscus will decrease until it reaches a value that satisfies the equation

$$h = \frac{2\sigma}{\rho g R},$$

where R is now not the radius of the tube but the radius of curvature of the meniscus, $r < R$.

2.55. Although the cross-sectional area of all four pipes of diameter $D/2$ each is equal to that of one pipe of radius D, the volume flow through these pipes is lower. This follows from Poiseuille's law

$$Q = \frac{\pi \Delta p D^4}{128 \eta l}.$$

Thus, the volume flow through each of the four pipes of $D/2$ diameter is lower than that through the big pipe not by a factor of four but by a factor of 16 (at the same pressure head), with the result that the total volume flow through the four pipes will be one-fourth of the flow through the big pipe.

2.56. The transverse distribution of velocities in the flow of a viscous liquid in a horizontal pipe is determined

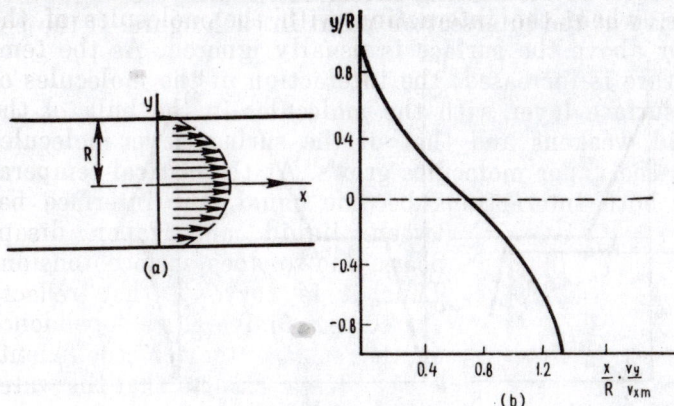

Fig. 2.56

via the formula (Figure (a))

$$v_x = v_{xm} \left[1 - \left(\frac{y}{R} \right)^2 \right].$$

The radial coordinate y is reckoned from the pipe's axis. The time it takes the particle to fall from the wall to a point whose ordinate is y is

$$t = (R - y)/v_y.$$

In the course of this fall the particle will be shifted in the horizontal direction over a distance

$$x = \int_R^y v_x \, dt = - \frac{v_{xm}}{v_y} \int_R^y \left[1 - \left(\frac{y}{R} \right)^2 \right] dy$$

$$= \frac{v_{xm}}{v_y} \left[\frac{2}{3} - \frac{y}{R} + \frac{1}{3} \left(\frac{y}{R} \right)^3 \right] R.$$

At the lowest possible point $(y = -R)$ we have

$$x = \frac{4}{3} \frac{v_{xm}}{v_y} R = \frac{2}{3} \frac{v_{xm}}{v_y} D.$$

183

The shape of the particle's trajectory in Figure (b) is represented in dimensionless coordinates, y/R and $(x/R) \times (v_y/v_{xm})$.

2.57. Each figure accompanying the problem contains the initial segments of the graphs representing the cooling of water or the heating of ice. Continuing these graphs, we arrive at the intersection point in each figure. If the point

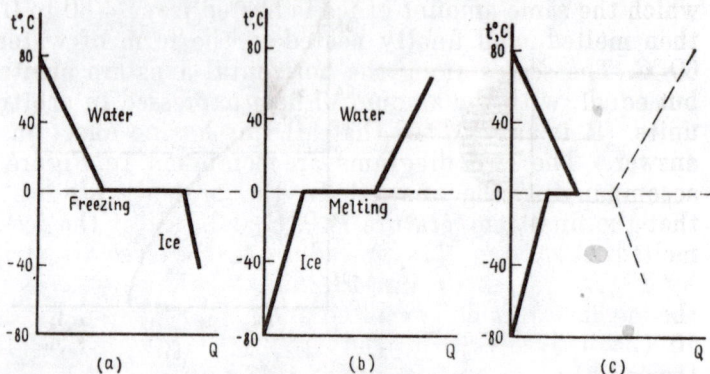

Fig. 2.57

of intersection lies above the horizontal line corresponding to a temperature of 0 °C, the final temperature is positive, when the point lies below this line, the final temperature is negative. If the graphs meet on the line $t = 0$ °C, the final temperature is 0 °C and the amount of the phase that has a horizontal section on the graph prior to intersection will decrease. The ratio of the length of this section to the total length of the horizontal section corresponding to this phase determines the fraction of the initial mass of this phase that has transformed into the other phase. When analyzing the graphs, we must bear in mind that the slopes of the straight lines are determined by the mass of water or ice and their specific heat capacity by the formula

$$\frac{\Delta t}{\Delta Q} = \frac{1}{cm}.$$

Here one must bear in mind that the specific heat capacity of water is twice as high as that of ice. The length of the horizontal sections corresponding to the water freezing or the ice melting is determined by the fact that the amount of heat required for melting a certain amount of

184

ice is equal to the amount of heat required for heating the same mass of water to 80 °C. For the sake of illustration, Figure (a) accompanying the answer shows the diagram for the cooling off of a mass of water from 80 to 0 °C, then the freezing of this water, and finally the cooling off of the ice down to $t = -40$ °C. Figure (b) accompanying the answer shows the reverse process in which the same amount of ice is heated from -80 to 0 °C, then melted, and finally heated in the form of water to 60 °C. The scales along the horizontal axes are arbitrary but equal, with the amount of heat expressed in arbitrary units. (It is easy to see that all this has no effect on the answer.) The two diagrams are combined in Figure (c) accompanying the answer. In the present case we see that the final temperature is 0 °C and half of the ice has melted. Applying this procedure to the case illustrated by Figure (a) accompanying the problem, we see that the ice has completely melted and the final temperature is 10 °C; for Figure (b) accompanying the problem, half of the ice has melted and the final temperature is 0 °C; for Figure (c) accompanying the problem, the case is similar to (b) but half of the water has frozen; finally, for Figure (d), all the water has frozen and the final temperature is -20 °C.

2.58. At the lowest possible pressures and the highest possible temperatures a substance may exist only in the vapor state (region *1*). Compressing the vapor at relatively high temperatures, we can transform it into the liquid state provided that the temperature is below the critical. The curve separating region *1* from region *2* corresponds to pressures and temperatures at which the liquid is in equilibrium with the saturated vapor of this liquid, with the region *2* corresponding to the liquid. T_{cr} on the temperature axis stands for the critical temperature. By cooling the liquid, we arrive at temperatures at which there is equilibrium between the liquid and the solid phase—this corresponds to region *3*. At low pressures there can be equilibrium between the vapor and the solid, but there is only one value of temperature and pressure at which equilibrium can exist between all three phases. This is the so-called triple point, and it is at this point that all three curves meet.

2.59. As distinct from the majority of substances, the ice-water system has an equilibrium curve with a negative slope. In view of this, higher pressures correspond to a lower temperature at which ice and water are in equilibrium. If ice was under an external pressure p_1 at a certain temperature and then this pressure was increased to p_3, then at a certain pressure p_2, whose value lies on the phase equilibrium curve, the ice will melt. The anomalous dependence of the melting point of ice on pressure is linked with the anomalous relation between the densities of water and ice. As a rule, the density of the solid phase is higher than that of the liquid, but for water the situation is the opposite: the density of ice is lower than that of water. This property is extremely important for the preservation of life in ponds, lakes, and rivers. If the density of water were lower than that of ice, all ponds, lakes and rivers would freeze solid.

2.60. Compressibility is defined thus:

$$\beta = -\frac{1}{V}\frac{dV}{dp},$$

whence $dV/V = -\beta\, dp$. Hence,

$$\int_{V_1}^{V_2}\frac{dV}{V} = -\int_{p_1}^{p_2}\beta\, dp.$$

Integration yields

$$\ln(V_1/V_2) = \int_{p_1}^{p_2}\beta\, dp.$$

The integral on the right-hand side gives the area bounded by the curve, the horizontal axis, and the vertical straight lines at p_1 and p_2. After evaluating this integral, we turn to the volume ratio. If the compressibility were pressure independent, the volume ratio would be

$$V_1/V_2 = \exp[\beta(p_2 - p_1)].$$

2.61. At a maximum point the derivative $d\rho/dt$ is zero. For this reason near a maximum the deviations in the density from the maximum value for small deviations in the precision with which the temperature is measured are at a minimum, with the result that in the neighborhood of the maximum the precision with which density is determined is the highest.

2.62. As is known, the heat flux is determined by the equation

$$\frac{dQ}{dt} = -\lambda \frac{dT}{dx} S.$$

Assuming that the heat flux is steady-state and, hence, dQ/dt is the same at all points of the wall, we find that where the absolute value of the gradient dT/dx is greater, the respective thermal conductivity coefficient is smaller. Hence, the inner layer of the wall has a higher thermal conductivity.

2.63. To elongate the rod by Δl, we must apply, according to Hooke's law, the force

$$F = \frac{ES}{l}\, \Delta l. \qquad (2.63.1)$$

The work of elongation performed from x to $x + \Delta x$ is

$$dA = F\, dx = \frac{ES}{l}\, x\, dx,$$

and the work performed from 0 to Δl is

$$A = \frac{ES}{2l}\, (\Delta l)^2.$$

Multiplying the numerator and denominator by l and introducing the notation

$$\Delta l / l = \varepsilon$$

(the strain, or extension per unit length), we get

$$A = \frac{ESl}{2}\, \varepsilon^2.$$

The performed work goes to increasing the internal energy of the rod, that is, the energy of elastic deformation. Dividing this energy by the volume of the rod, we get the bulk energy density

$$w = E\varepsilon^2/2.$$

From (2.63.1) it follows that

$$E\varepsilon = F/S = \sigma,$$

where σ is the internal mechanical stress. For this reason, the bulk energy density can be represented as

$$w = \sigma\varepsilon/2.$$

2.64. For each bar the thermal linear strain is

$$\Delta l/l = \alpha \Delta T,$$

while the mechanical linear strain is

$$\Delta l/l = -\sigma/E,$$

where σ is the internal mechanical normal stress (Young's modulus), which is the same for both bars. The sum of the two strains is zero:

$$\alpha \Delta T - \sigma/E = 0.$$

Hence, $\alpha E = \sigma/\Delta T$. Since the right-hand side is the same for both bars, we can write $\alpha_1 E_1 = \alpha_2 E_2$, or

$$\alpha_1/\alpha_2 = E_2/E_1.$$

If the walls possess the same mechanical properties, the deformability of the walls has no effect on the result.

3. Electrostatics

3.1. The components of the electric field strength that are generated by the charges at the acute angles are equal and are directed toward the negative charge. If we denote the length of the hypotenuse by $2a$, each of these components is $Q/4\pi\varepsilon_0\varepsilon a^2$ and the sum is $Q/2\pi\varepsilon_0\varepsilon a^2$. The component of the electric field strength generated by the charge $+2Q$ is the same. It is directed at right angles to the hypotenuse away from the right angle. The resultant field strength is directed parallel to the leg connecting the charges $+2Q$ and $-Q$ along vector 3.

3.2. Since in the case at hand all the electric field vectors lie on a single straight line, the vector sum may be replaced with the scalar sum. For unlike charges the direction of the resultant vector does not change while for like charges it does. In the case illustrated by Figure (a), the electric field strength is positive everywhere. Allowing for the signs specified in the problem, we conclude that the left charge is positive and the right charge is negative. Similarly, for the case illustrated by Figure (c), the left charge is negative and the right charge is positive. In Figures (b) and (d) the electric field strength changes its sign at the midpoint of the distance between the charges. Obviously, this can only occur if the charges are like. Bearing in mind the aforesaid and allowing for the relationship between the direction of the electric field vector and the sign of the charge generating the field, we conclude that for the case depicted in Figure (b) both charges are positive, while for the case depicted in Figure (d) both charges are negative.

3.3. Since both electric field vectors lie on a single straight line, they can be added algebraically, just as we did in the previous problem. The electric field strength to the right of charge Q_b in the immediate vicinity of the charge is negative; hence, the charge is negative (the electric field vector is directed toward the charge). The electric field strength may be positive to the right of Q_b only if Q_a is positive and greater (in absolute value) than Q_b. The electric field strength is zero at point x_1 if

$$\frac{Q_a}{(l+x_1)^2} - \frac{Q_b}{x_1^2} = 0,$$

whence

$$\frac{Q_a}{Q_b} = \left(\frac{l+x_1}{x_1}\right)^2.$$

At all points that are to the right of Q_b the electric field strength is specified by the equation

$$E_x = \frac{Q_a}{(l+x)^2} - \frac{Q_b}{x^2}.$$

Taking the derivative with respect to x and nullifying it, we find that the maximum is at the point

$$x_2 = \frac{l}{(Q_a/Q_b)^{1/3} - 1}.$$

3.4. The direction of the electric field vector at a point with coordinates x and y (see the figure accompanying the answer) is determined by the two components, E_x and E_y:

$$E_x = \frac{\tau_2}{2\pi\varepsilon_0 x}, \quad E_y = \frac{\tau_1}{2\pi\varepsilon_0 y}.$$

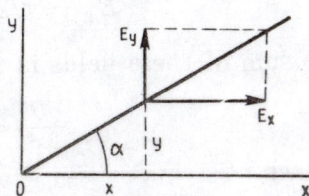

Fig. 3.4

For the extension of the resultant vector to pass through the origin, which is where the conductors intersect, the slope of the vector must be equal to y/x, that is,

$$\frac{E_y}{E_x} = \frac{\tau_1 x}{\tau_2 y} = \frac{y}{x}.$$

Thus,

$$\tan \alpha = y/x = \sqrt{\tau_1/\tau_2}.$$

3.5. No such point can exist in region *II*, since the electric field vectors of the two charges point in the same direction—from the linear charge to the point charge. In regions *III* and *I* the electric field vectors of these charges

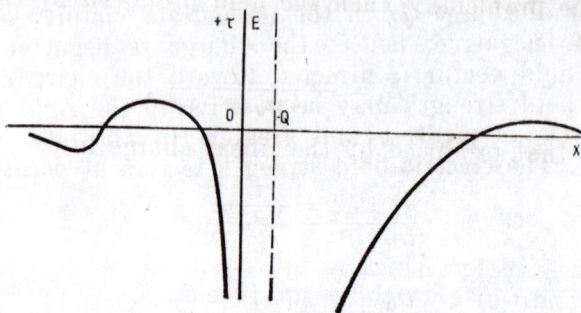

Fig. 3.5

point in different directions. Let us examine each region separately. At a certain point to the right of the point charge, the electric field strength produced by this charge is

$$E_1 = -Q/4\pi\varepsilon_0 x^2,$$

where x is the distance from the charge to the point. The linear charge produces the following field at the same point:

$$E_2 = \tau/2\pi\varepsilon_0 (x + a).$$

The sum of these fields is zero if

$$\frac{Q}{2x^2} = \frac{\tau}{a+x},$$

whence

$$x = \frac{Q}{4\tau} \pm \sqrt{\frac{Q^2}{16\tau^2} + \frac{aQ}{2\tau}}.$$

Only the plus sign in front of the radical sign has any meaning, since the minus sign corresponds to a point to the left of the point charge, where the electric field strengths of both charges are added rather than subtracted from each other (the quantities are equal in absolute value). Now let us turn to region I, that is, to the left of the linear charge. To see whether there are points in this region where the electric field strength is zero, we determine the electric field strengths produced by the two charges in this region. For the sake of convenience we direct the x axis to the left and take point A on the linear conductor as the origin (see the figure accompanying the problem). Then the field produced by the point charge is

$$E_1 = - \frac{Q}{4\pi\varepsilon_0 (a+x)^2},$$

while that produced by the linear charge is

$$E_2 = \frac{\tau}{2\pi\varepsilon_0 x}.$$

The two vectors point in opposite directions, obviously. The condition that their sum is zero yields the following equation for x:

$$x^2 + \left(2a - \frac{Q}{2\tau}\right) x + a^2 = 0,$$

whence

$$x = \frac{1}{2}\left(\frac{Q}{2\tau} - 2a\right) \pm \sqrt{\frac{1}{4}\left(\frac{Q}{2\tau} - 2a\right)^2 - a^2}.$$

The net field strength in region I is zero if the radicand is positive, obviously, that is, if

$$Q \geqslant 8a\tau.$$

If this condition is met, region I contains two points where the electric field is zero. The distribution of the electric field strength along the x axis is shown schematically (without a definite scale) in the figure accompanying the answer.

3.6. Let us first solve this problem by dimensional considerations. Here are the quantities on which the interaction force between the conductors might depend: the charge densities, the distance between the conductors, and the "absolute" permittivity

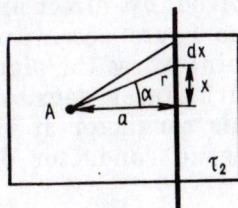

Fig. 3.6

$$\varepsilon_a = \varepsilon_0 \varepsilon,$$

191

which obviously has the same dimensions as the permittivity of empty space ε_0, since the dielectric constant ε is dimensionless. The SI dimensions of these quantities are

$$[F] = LMT^{-2}, \quad [\tau] = L^{-1}TI, \quad [\varepsilon_a] = L^{-3}M^{-1}T^4I^2,$$
$$[a] = L.$$

Assuming that these quantities enter the expression for force F with exponents p, q, and r, we can write

$$F = C\tau^p\varepsilon_a^q a^r$$

(C is a dimensionless constant), and the equation for the dimensions is

$$LMT^{-2} = [L^{-1}TI]^p \times [L^{-3}M^{-1}T^4I^2]^q \times L^r.$$

This yields the following equations for the exponents:

$$1 = -p - 3q + r, \quad 1 = -q, \quad -2 = p + 4q,$$
$$0 = p + 2q.$$

Hence,

$$p = 2, \quad q = -1, \quad r = 0,$$

or

$$F = C\frac{\tau_1\tau_2}{\varepsilon_0\varepsilon}. \tag{3.6.1}$$

We have found, therefore, that the interaction does not depend on the distance between the conductors.

It goes without saying that C cannot be determined by dimensional analysis alone. The same problem can be solved by direct integration via the Coulomb law. In the figure accompanying the answer, A stands for the point where the plane of the drawing "cuts" the conductor with linear density τ_1. The electric field generated by this conductor at the point with the element dx of the second conductor distant r from the first is

$$E = \frac{\tau_1}{2\pi\varepsilon_0\varepsilon r}.$$

The following force acts on element dx of the second conductor:

$$dF = E\tau_2 \, dx.$$

We are interested, however, in the component of the force that is perpendicular to the second conductor, or $dF \cos \alpha$, since the longitudinal component is canceled out by an equal component acting on the symmetrical element. Let us express all linear quantities in terms of distance a and angle α:

$$r = \frac{a}{\cos \alpha}, \qquad dx = \frac{a}{\cos^2 \alpha} \, d\alpha.$$

Substituting these quantities into the expression for the perpendicular component of the force acting on element dx, we get (after canceling out like terms)

$$dF = \frac{\tau_1 \tau_2}{2\pi \varepsilon_0 \varepsilon} \, d\alpha.$$

Integration from $-\pi/2$ to $+\pi/2$ yields

$$F = \frac{\tau_1 \tau_2}{2\varepsilon_0 \varepsilon},$$

that is, we arrive at an expression of the (3.6.1) type. Hence $C = 1/2$.

3.7. The element of the disk bounded by radii ρ and $\rho + d\rho$ and angle $d\varphi$ carries a charge (taking into account both sides of the disk) equal to $2\sigma\rho \, d\rho \, d\varphi$. At a distance z from this element and, hence, at a distance r from the disk's center (Figure (a)), the electric field generated by this charge is

$$E = \frac{2\sigma\rho \, d\rho \, d\varphi}{4\pi\varepsilon_0 \varepsilon z^2}.$$

Only the component of this field that points in the direction of r is of any interest to us since the perpendicular component is canceled out by an equal component (pointing in the opposite direction) from the symmetrically situated charge. For this reason, the charge on the disk limited by the radii ρ and $\rho + d\rho$ creates an electric field

$$dE = \frac{\rho \, d\rho \, \sigma \cos \alpha}{z^2}. \qquad (3.7.1)$$

We express all geometric quantities in terms of distance r and angle α:

$$z = \frac{r}{\cos \alpha}, \qquad \rho = r \tan \alpha, \qquad d\rho = \frac{r \, d\alpha}{\cos^2 \alpha}.$$

After substituting into (3.7.1) and canceling out like terms, we get

$$dE = \frac{\sigma \sin \alpha \, d\alpha}{\varepsilon_0 \varepsilon}.$$

Integration from $\alpha = 0$ to the value α_m corresponding to the edge of the disk yields

$$E = \frac{\sigma}{\varepsilon_0 \varepsilon} (1 - \cos \alpha_m) = \frac{\sigma}{\varepsilon_0 \varepsilon} \left(1 - \frac{r}{\sqrt{R^2 + r^2}} \right). \quad (3.7.2)$$

For $r \ll R$, angle α is close to 90°. In this case, $E \approx \sigma / \varepsilon_0 \varepsilon$, just as in the case with an infinitely large plate.

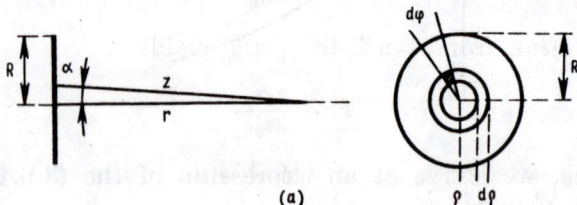

Fig. 3.7a

Let us calculate E for $r \gg R$. To this end we express $\cos \alpha_m$ in terms of r and R:

$$\cos \alpha_m = \frac{r}{\sqrt{R^2 + r^2}}.$$

Using the rules of approximate calculations, we arrive at

$$1 - \frac{r}{\sqrt{R^2 + r^2}} = 1 - \frac{1}{\sqrt{1 + R^2/r^2}} \approx \frac{R}{2r^2}.$$

Substituting this into (3.7.2), we get

$$E = \frac{\sigma R^2}{2\varepsilon_0 \varepsilon r^2}.$$

Since $\sigma = Q/2\pi R^2$, we have

$$E = \frac{Q}{4\pi \varepsilon_0 \varepsilon r^2},$$

just as for a point charge (see the problem).

Figure (b) shows the variation of the electric field of the disk with distance (curve 1); for comparison, the straight line *3* corresponds to the field created by an infinitely large plate with a surface charge density equal to that of the disk, while curve *2* corresponds to the field of a point charge whose magnitude coincides with the charge of the disk. Dimensionless coordinates are

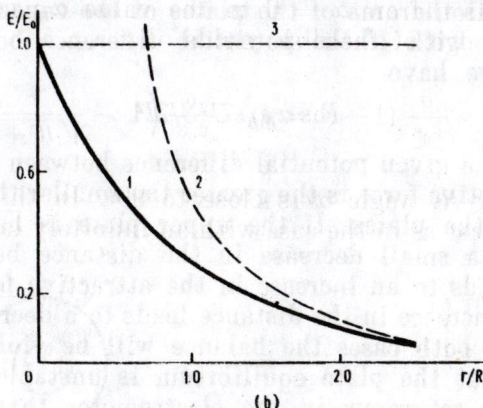

(b)

Fig. 3.7b

employed in Figure (b): r/R along the horizontal axis and E/E_0 along the vertical axis (E_0 is the electric field strength generated by the infinitely large plate).

3.8. The force with which an electric field acts on a dipole is

$$F = p_{el} \frac{dE}{dr}. \qquad (3.8.1)$$

Since an infinitely long straight conductor with an evenly distributed charge (density) generates an electric field

$$E = \frac{\tau}{2\pi\varepsilon_0 \varepsilon r},$$

we have (according to (3.8.1))

$$F = -\frac{\tau p_{el}}{2\pi\varepsilon_0 \varepsilon r^2}. \qquad (3.8.2)$$

Nothing was said in the problem about the sign of the charge on the conductor. Obviously, if the charge is positive and the dipole moment coincides in direction with the positive direction of the electric field vector, the dipole will move toward the conductor, which agrees with the "minus" sign in (3.8.2).

3.9. If the field in the region between the plates can be assumed to be uniform, the plates of the parallel-plate capacitor interact with a force

$$F = \varepsilon_0 \varepsilon E^2 S / 2,$$

where S is the area of the plates of the capacitor. Since $E = U/l$, with U the potential difference between the plates, we have

$$F = \varepsilon_0 \varepsilon U^2 S / 2l^2.$$

Thus, for a given potential difference between the plates, the attractive force is the greater the smaller the distance between the plates. If the upper plate is balanced by weights, a small decrease in the distance between the plates leads to an increase in the attractive force, while a small increase in the distance leads to a decrease in the force. In both cases the balance will be violated. This means that the plate equilibrium is unstable. There is a special set screw in the electrometer that does not allow the upper plate to move below the level at which the measurement is taken.

3.10. The force acting on the strip when the strip lies on the lower plate is determined by the formula for the attractive force between the plates of a parallel-plate capacitor,

$$F = \frac{\varepsilon_0 \varepsilon E^2}{2} S = \frac{QE}{2},$$

where S is the area of the strip, as if it was part of the lower plate of the capacitor. When this force becomes greater than the weight of the strip, the strip begins to move upward, but retains its charge $Q = \sigma S$.

When the distance between the strip and the lower plate becomes great, the strip will not only be attracted by the upper plate but will also be repulsed by the lower plate where the charge density will gradually become even. As a result, the force on the strip increases in magnitude. If we ignore the distortions introduced by the charge of the strip into the field (this can be done if the strip is small), we can assume that the strip is in a field of strength E and that the following force acts on it: $F = QE$. The charged strip induces a charge on the upper plate as it approaches the plate. This leads to a distortion in the field and a slight increase in F. Although in the above discussion we have considered a flat strip, the same line of reasoning is valid qualitatively for any small conductor lying, at the initial stage, on the lower plate of the capacitor.

3.11. Let us first solve this problem by dimensional analysis. The following quantities are present in the problem: the initial potential difference U that the electron or ion has to pass, the potential difference U_0 between the plates, the distance d between the plates, the sought distance l that the electron or ion has to travel before it hits the plate, the charge Q of the particle, and the particle's mass m. The equation for the dimensions can be written as follows:

$$[l] = [d]^a[U_0]^b[U]^c[Q]^x[m]^y,$$

or

$$L = L^a[L^2 M T^{-3} I^{-1}]^{b+c}[IT]^x M^y.$$

For the exponents we have the following four equations:

$$a + 2b + 2c = 1, \; b + c + y = 0,$$
$$x - 3b - 3c = 0, \; x - b - c = 0,$$

whence

$$a = 1, \quad b = -c, \quad x = 0, \quad y = 0.$$

We see that the distance traveled by the particle (an electron or an ion) does not depend on the charge-to-mass ratio.

We arrive at the same result if we solve the equation of motion of the particle. Under the potential difference U_0, the particle acquires a velocity

$$v = \sqrt{2QU_0/M},$$

with which it moves parallel to the plates, while the acceleration with which the particle moves transversely to the plates is

$$w = QU/md.$$

The particle takes a time interval

$$t = \sqrt{\dfrac{d}{w}} = d\sqrt{\dfrac{m}{QU}}$$

to cover the distance

$$l = vt = d\sqrt{2U_0/U}.$$

This conclusion has a broader meaning than the one obtained earlier. It follows that for a given initial energy, a charged particle moves in an electric field along a trajectory that does not depend on the particle's charge-to-mass ratio.

3.12. A dipole that is placed in a nonuniform electric field and is oriented along the field's direction is under a force

$$F = p_{el} \frac{dE}{dr},$$

where p_{el} is the dipole electric moment. If the direction of the dipole's axis is taken as the positive direction, the direction of the force will be determined by the sign of the derivative. In the case at hand the derivative is negative and, hence, the dipole is moving toward the point charge.

3.13. A point dipole oriented along the lines of force of the field created by a point charge is under a force

$$F_p = p_{el} \frac{dE_Q}{dr}.$$

Since the electric field created by a point charge is

$$E_Q = \frac{Q}{4\pi\varepsilon_0\varepsilon r^2},$$

we can write

$$\frac{dE_Q}{dr} = -\frac{Q}{2\pi\varepsilon_0\varepsilon r^3},$$

with the result that the force acting on the dipole is

$$F_p = -\frac{Qp_{el}}{2\pi\varepsilon_0\varepsilon r^3}.$$

At points that lie on the axis of the point dipole, the electric field of the dipole is

$$E_p = \frac{p_{el}}{2\pi\varepsilon_0\varepsilon r^3}.$$

When a point charge Q is in this field, the force acting on it is

$$F_Q = \frac{Qp_{el}}{2\pi\varepsilon_0\varepsilon r^3}.$$

In accordance with Newton's third law, this force must coincide in magnitude with, but be opposite to, force F_p.

The positive direction in the figure accompanying the problem is the one from the point charge to the dipole. Therefore, the "minus" sign in the force acting on the dipole implies that this force is directed toward the point charge. The field created by the dipole at the point where the point charge is positioned has a "plus" sign, that is, is directed toward the dipole. The force acting on the point charge points in the same direction.

3.14. The electric field in which the sphere is placed induces charges of opposite sign on the sphere, in view of which the sphere becomes a dipole. After the sphere is shifted, it finds itself in a nonuniform field, which forces it to move toward the charge to which it was shifted. Thus, the equilibrium of the sphere at the midpoint between the charges is unstable.

3.15. Due to electrostatic induction, one side of the sphere becomes positively charged, while the other becomes negatively charged, and the sphere becomes a dipole. At first glance it might seem that since the dipole is oriented along the lines of force of the field and the field of the capacitor is uniform, no forces act on the sphere. But this is not so. The presence of the sphere will distort the field. The charge density, and hence the field strength, at the points of the plates that lie on the straight line that is perpendicular to the plates and passes through the center of the sphere will increase. The dipole will find itself in a nonuniform field and will be attracted to the plate that is closer to it. If the string enables the sphere to touch the plate, the sphere will lose its charge, which is opposite to the one on the plate. But the sphere will then acquire a charge that is of the same sign as that on the plate it has just touched. This leads to a repulsive force between sphere and plate, with the result that the sphere will move toward the other plate. After touching this plate (if the string enables it to do this), the sphere will reverse the sign of its charge and will move in the direction of the first plate, and so on.

3.16. If the distance between the spheres is not very large, the charges on the spheres are not evenly distributed over the surfaces. The effect of the spheres on each other results in that in the case of like charges the sections of the spheres that are farthest from each other will have an enhanced charge density, while in the case of unlike charges the sections of the spheres that are closest to each other will have an enhanced charge density. For this reason, the distance between the "centers of charge" for like charges is greater than that for unlike charges. Hence, the attractive force between the unlike charges will be greater (in magnitude) than the repulsive force between the like charges.

3.17. The field strength in each layer is

$$E = \frac{Q}{4\pi\varepsilon_0 \varepsilon r^2}.$$

On the log-log scale,

$$\log E_1 = \log \frac{Q}{4\pi\varepsilon_0} - \log \varepsilon_1 - 2 \log r_1 \qquad (3.17.1)$$

and

$$\log E_2 = \log \frac{Q}{4\pi\varepsilon_0} - \log \varepsilon_2 - 2 \log r_1 \qquad (3.17.2)$$

in each layer at the boundary between the layers. Subtracting (3.17.1) from (3.17.2) and bearing in mind that the difference of the logarithms of two quantities equals the logarithm of the ratio of these quantities, we have

$$\log (E_2/E_1) = \log (\varepsilon_1/\varepsilon_2).$$

Hence, in the inner layer the dielectric constant is higher than in the outer. The difference of the logarithms of the field strengths in Figure (b) accompanying the problem is about 0.3, which corresponds to the ratio of the dielectric constants of about 2.

3.18. The lines of force of electric induction become denser as one moves closer to the solid dielectric, which means that the density of bound charges on the surface of the solid dielectric becomes enhanced. This density is the higher the greater the dielectric constant. Whence $\varepsilon_2 > \varepsilon_1$.

3.19. The potential at each point is the algebraic sum of potentials of the field of each charge. For a point charge, the potential at distance r from the charge is

$$\varphi = \frac{Q}{4\pi\varepsilon_0 \varepsilon r}$$

(it is assumed that the potential at infinity is zero). When the charges are like, the absolute value of the potential at a point r distant from one of the charges is

$$\varphi = \frac{Q}{4\pi\varepsilon_0 \varepsilon} \left(\frac{1}{r} + \frac{1}{l-r} \right).$$

The sign of the potential coincides with that of the charge. Hence, in Figure (a) both charges are positive, while in Figure (c) both are negative. When the charges are unlike, the potential at midpoint between the charges is zero. The potential is positive closer to the positive charge to the left in the case shown in Figure (b) and to the right in the case shown in Figure (d).

3.20. The field strength vanishes only at one point, 3, where the derivative $d\varphi/dr$ is zero. Since near charge Q_2

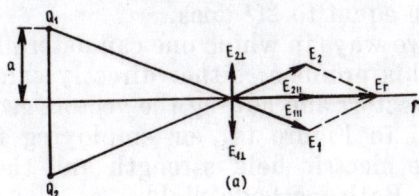

(a)

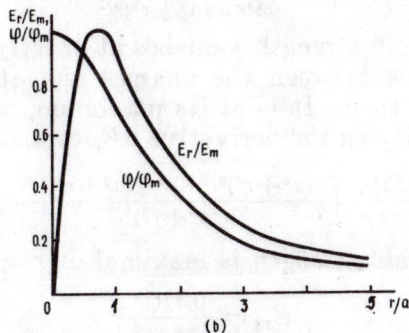

(b)

Fig. 3.21

the potential is negative while near Q_1 it is positive, we can conclude that Q_2 and Q_1 are negative and positive, respectively. The potential at every point in space is the algebraic sum of the potentials produced by all charges. To the right of Q_2 (except in the immediate vicinity of Q_2) the potential is positive. This implies that in the entire region to the right of Q_2 the potential produced by Q_1 is greater in absolute value than the potential produced by Q_2. Hence, the absolute value of Q_1 is greater than that of Q_2, too.

3.21. Since potentials must be added algebraically, we conclude that at a point removed from the middle of the distance between the charges by an interval of r the potential is

$$\varphi = \frac{Q}{2\pi\varepsilon_0\varepsilon\,(a^2+r^2)^{1/2}}$$

(the potential at infinity is assumed to be equal to zero). Hence, the potential falls off as r increases in exactly the same manner on both sides of the straight line connecting the charges. At great distances $(r \gg a)$, φ varies in exactly the same way as the potential produced by a point charge equal to $2Q$ does.

There are two ways in which one can determine the electric field in this problem: either directly calculating the values of the vectors and adding the vectors geometrically, just as shown in Figure (a), or employing the formula that links the electric field strength and the potential, $E = -d\varphi/dr$. Both methods yield

$$E_r = \frac{Qr}{2\Omega\varepsilon_0\varepsilon\,(a^2+r^2)^{3/2}}\ .$$

The electric field strength vanishes at exactly the middle of the distance between the charges and at an infinite distance from them. It is at its maximum, which can be found by nullifying the derivative dE_r/dr:

$$\frac{dE_r}{dr} = \frac{2Q}{\pi\varepsilon_0\varepsilon}\left[\frac{(a^2+r^2)^{3/2}-3r^2\,(a^2+r^2)^{1/2}}{(a^2+r^2)^{5/2}}\right]=0.$$

The electric field strength is maximal at $r = a/\sqrt{2}$, with

$$E_\mathrm{m} = \frac{0.77Q}{4\pi\varepsilon_0\varepsilon a^2}\ .$$

Figure (b) shows the behavior of E and φ in dimensionless coordinates: $\varphi/\varphi_\mathrm{m}$, E/E_m, and r/a.

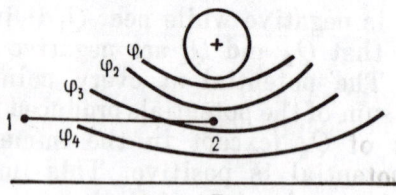

Fig. 3.22

3.22. All the equipotential surfaces of the field between the sphere and the plate are convex downward (that is, toward the plate). Hence, on any straight line parallel to the plate, the points farther from the sphere have a potential lower than those closer to the sphere. Hence, the point charge is moved from a point with a lower potential to a point with a higher potential. This requires doing work against the forces of the electric field.

3.23. Point *1* has a positive potential with respect to the negatively charged plate of *C1*. This potential is half the difference in potential between the plates of *C1* (and of *C2*). Since point *2* lies in capacitor *C2* closer to the negatively charged plate, its potential is lower than that at point *1*. When the point charge is moved from point *1* with a higher potential to point *2* with a lower potential, the electric field performs work equal to the product of the strength of the point charge by the potential difference between points *1* and *2*:

$$A = Q(\varphi_1 - \varphi_2) > 0.$$

3.24. Initially the capacitance of the capacitor (filled with the dielectric) is $C = \varepsilon_0 \varepsilon ab/l$. After the dielectric is moved out of the capacitor by a distance x, the capacitance becomes

$$C = \varepsilon_0 a [x + \varepsilon(b - x)]/l.$$

Since the total charge on the plates of the capacitor remains unchanged, the potential difference between the plates becomes

$$U = \frac{Ql}{\varepsilon_0 a [x + \varepsilon(b - x)]},$$

where Q is the charge on the plates. Since initially the potential difference was $U = Ql/\varepsilon_0 \varepsilon ab$, we have

$$\frac{U}{U_0} = \frac{\varepsilon b}{x + \varepsilon(b - x)} = \frac{\varepsilon b}{\varepsilon b - (\varepsilon - 1)x}.$$

The field strength between the plates will increase by the same factor. The charge density in the part without the dielectric is

$$\sigma_1 \varepsilon = {}_0 E = \frac{\varepsilon_0 Q}{a[\varepsilon b - (\varepsilon - 1)x]},$$

while on the part with the dielectric it is

$$\sigma_2 = \varepsilon_0 \varepsilon E = \frac{\varepsilon_0 \varepsilon Q}{a[\varepsilon b - (\varepsilon - 1)x]}.$$

Initially the charge density on each plate was

$$\sigma_0 = Q/ab,$$

or, respectively

$$\frac{\sigma_1}{\sigma_0} = \frac{b}{\varepsilon b - (\varepsilon - 1) \, x/b} \quad \text{and} \quad \frac{\sigma_2}{\sigma_0} = \frac{\varepsilon}{\varepsilon - (\varepsilon - 1) \, x/b} \, .$$

In the part filled with the dielectric, the charge density gradually grows in the same proportion as the electric field strength and the potential difference between the plates, while the total charge of this part gradually decreases due to the increase in x. In the part not filled with the dielectric, the charge density first drops ε-fold (at $x \ll b$) and then gradually grows, approaching the value it had when the dielectric filled the entire space between the plates.

3.25. Being a conductor, each plate has the same potential at each point, while the electric field strength, which is minus one multiplied by the gradient of the potential, is highest where the plates are closest to each other. At the same time, the electric field strength near the surface of a conductor is linked with the local surface charge density through the formula $E = \sigma/\varepsilon_0\varepsilon$. For this reason, the surface charge density at point *1* is higher than that at point *2*.

3.26. The electric field strength at the core is

$$E_1 = \frac{2U}{D_1 \ln (D_2/D_1)} \, .$$

To find the extremum of E_1 we take the derivative,

$$\frac{dE_1}{dD_1} = -2U \, \frac{(\ln D_2 - \ln D_1) - 1}{[D_1 \, (\ln D_2 - \ln D_1)]^2} \, ,$$

and nullify it. The result is

$$\ln D_2 - \ln D_1 = 1,$$

or

$$D_1 = D_2/e.$$

This corresponds to a minimum, since E_1 tends to ∞ as $D_1 \to 0$ and $D_1 \to D_2$.

3.27. Since the charges on the capacitors *C1* and *C2* are equal, the potential difference across these capacitors and the capacitance of each capacitor are linked through the following formula:

$$C_1 U_1 = C_2 U_2. \qquad (3.27.1)$$

For capacitors *C3* and *C4* there is a similar formula:

$$C_3 U_3 = C_4 U_4. \qquad (3.27.2)$$

For a potential difference between points a and b to be zero, we must make sure that $U_1 = U_3$ and $U_2 = U_4$. Dividing (3.27.1) by (3.27.2) termwise and canceling equal potential differences, we get

$$C_1/C_3 = C_2/C_4.$$

Note that if a constant potential difference is applied between points A and B and the capacitors leak some charge (i.e. their resistance is not very high), the distribution of potential between the capacitors is the same as in the Wheatstone bridge, that is, is proportional to the resistances.*

* These considerations must be taken into account in some other problems, too (e.g. see Problems 3.30 and 3.31).

3.28. The charge of the solid sphere is

$$Q = \frac{4}{3} \pi \rho R^3,$$

where ρ is the volume charge density. Outside the sphere, that is, for $r > R$, the electric field strength coincides with the electric field strength of the same charge Q concentrated, however, at the center of the sphere:

$$E = \frac{Q}{4\pi\varepsilon_0 \varepsilon r^2} = \frac{1}{3} \frac{\rho R^3}{\varepsilon_0 \varepsilon r^2} . \qquad (3.28.1)$$

On the surface of the sphere,

$$E_R = \frac{\rho R}{3\varepsilon_0 \varepsilon} . \qquad (3.28.2)$$

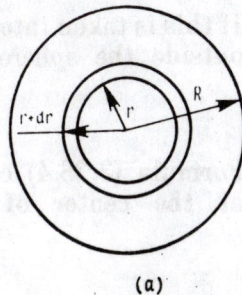

(a)

Fig. 3.28a

To find the electric field inside the sphere, we isolate a sphere of radius $r < R$ inside the sphere (Figure (a) accompanying the answer). The charge contained in this smaller sphere is $4\pi\rho r^3/3$. According to Gauss's theorem, the electric field at the boundary of the isolated sphere is

$$E = \frac{4\pi\rho r^3}{3 \times 4\pi\varepsilon_0\varepsilon r^2} = \frac{1}{3}\frac{\rho r}{\varepsilon_0\varepsilon}. \tag{3.28.3}$$

Thus, the electric field along r behaves in two ways: inside the sphere it increases linearly with r according to (3.28.3) from zero to the value given by formula (3.28.2), while outside the sphere it decreases by a quadratic (hyperbolic) law, just as in the case of a point charge.

The behavior of the potential inside and outside the sphere must also be considered separately. Inside the sphere,

$$\int_{\varphi_0}^{\varphi} d\varphi = -\frac{\rho}{3\varepsilon_0\varepsilon}\int_0^r r\,dr = -\frac{\rho}{6\varepsilon_0\varepsilon}r^2, \quad \varphi = \varphi_0 - \frac{1}{6}\frac{\rho r^2}{\varepsilon_0\varepsilon}.$$

At the boundary of the sphere,

$$\varphi_R = \varphi_0 - \frac{1}{6}\frac{\rho R^2}{\varepsilon_0\varepsilon}.$$

Finally, outside the sphere the potential is distributed thus:

$$\int_{\varphi_R}^{\varphi} d\varphi = -\frac{1}{3}\frac{\rho R^3}{\varepsilon_0\varepsilon}\int_R^r \frac{dr}{r^2}, \quad \varphi - \varphi_R = \frac{1}{3}\frac{\rho R^3}{\varepsilon_0\varepsilon}\left(\frac{1}{r} - \frac{1}{R}\right).$$

Putting $\varphi = 0$ at $r = \infty$, we get

$$\varphi_R = \frac{1}{3}\frac{\rho R^2}{\varepsilon_0\varepsilon}. \tag{3.28.4}$$

If this is taken into account, we can write for the potential outside the sphere the following formula:

$$\varphi = \frac{1}{3}\frac{\rho R^3}{\varepsilon_0\varepsilon r}.$$

Formula (3.28.4) can also be used to find the potential at the center of the sphere:

$$\varphi_0 = \frac{1}{2}\frac{\rho R^2}{\varepsilon_0\varepsilon}.$$

For the potential distribution inside the sphere we then get

$$\varphi = \frac{\rho}{2\varepsilon_0\varepsilon}\left(R^2 - \frac{r^2}{3}\right).$$

Figure (b) accompanying the answer shows the behavior of the electric field and the potential inside and outside

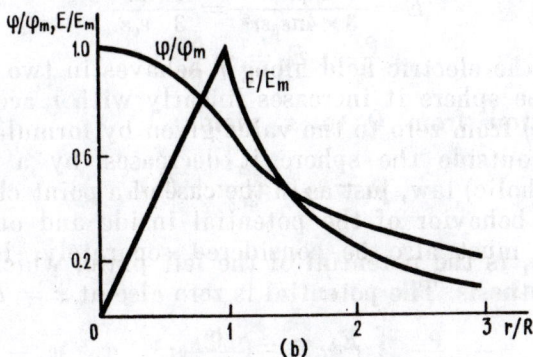

Fig. 3.28b

the sphere. Dimensionless coordinates φ/φ_m, E/E_m, and r/R are employed.

3.29. Let us isolate a thin layer of thickness dx parallel to the plates and lying between them (Figure (a) accompanying the answer). A unit area of this layer carries a volume charge ρdx. According to Gauss's theorem, the electric field generated by this layer is equal in absolute value (on each side of the layer) to

$$dE^* = \rho dx/2\varepsilon_0.$$

If all the charges to the left of the isolated layer generate a field of strength

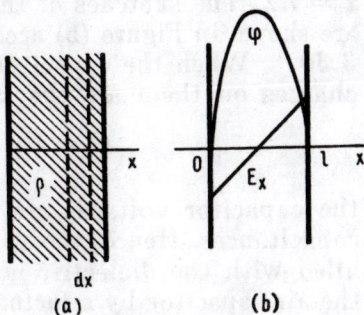

Fig. 3.29

E, the resultant electric field strength is $E - dE^*$ at the left boundary of the layer and $E + dE^*$ at the right. Thus, over a distance of dx the electric field strength increases by

$$dE = 2dE^* = \rho \, (dx/\varepsilon_0).$$

Integration yields

$$E = \rho x/\varepsilon_0 + E_0,$$

207

with E_0 the electric field at the left plate. According to the basic equation of electrostatics (the one that links the electric field strength with the potential),

$$\frac{\rho x}{\varepsilon_0} + E_0 = -\frac{d\varphi}{dx}.$$

Integration from 0 to x yields

$$\varphi_1 - \varphi_2 = \frac{\rho x^2}{2\varepsilon_0} + E_0 x, \qquad (3.29.1)$$

where φ_1 is the potential of the left plate, which is zero by hypothesis. The potential is zero also at $x = l$. Hence,

$$E_0 = -\rho l/2\varepsilon_0.$$

Substituting this into (3.29.1), we arrive at the relationship between φ and x:

$$\varphi = \frac{\rho}{2\varepsilon_0} x (l - x).$$

This function represents a parabola with a maximum at $x = l/2$. The sketches of the φ vs. x and E vs. x curves are shown in Figure (b) accompanying the answer.

3.30. When the capacitors are connected in series, the charges on them are the same. Since these charges are

$$Q = C_1 U_1 = C_2 U_2,$$

the capacitor voltages are inversely proportional to the capacitances. Hence, the voltage applied to the capacitor filled with the dielectric is smaller than that applied to the air capacitor by a factor equal to the ratio of the dielectric constant to unity (the dielectric constant of air, roughly).

3.31. If C_0 is the initial capacitance of each capacitor, the total initial capacitance of the two capacitors is $C = C_0/2$. After the distance between the plates of one capacitor is increased, the capacitance of this capacitor, C', becomes smaller than C_0. The voltage U_0 applied to the capacitors is distributed among the capacitors in inverse proportion to the capacitances, since the charge on the plates is

$$Q = U_1 C_0 = U_2 C'$$

Since U_0 remains unchanged, the voltage across the capacitor whose plates are not moved will decrease, while that across the second capacitor will increase.

If the capacitors are first charged and then disconnected from the DC source, the charge on them will remain unchanged. The voltage across each capacitor will be

$$U_1 = Q/C_0, \quad U_2 = Q/C_1.$$

For this reason, the potential difference across the capacitor whose plates are not moved remains unchanged, while that across the second capacitor increases.

3.32. When the capacitors are connected in parallel, the initial capacitance is $C = 2\varepsilon_0\varepsilon S/l$. After the distance between the plates is changed, the capacitance becomes

$$C = \frac{\varepsilon_0\varepsilon S}{l+a} + \frac{\varepsilon_0\varepsilon S}{l-a} = \frac{2\varepsilon_0\varepsilon S}{l - a^2/l}.$$

The new capacitance is greater than the initial one.

When the capacitors are connected in series,

$$\frac{1}{C} = \frac{2l}{\varepsilon_0\varepsilon S}.$$

After the distance between the plates is changed,

$$\frac{1}{C} = \frac{l+a}{\varepsilon_0\varepsilon S} + \frac{l-a}{\varepsilon_0\varepsilon S} = \frac{2l}{\varepsilon_0\varepsilon S},$$

that is, the capacitance remains unchanged.

3.33. The electric displacement vector has the same length in both halves, and since $E = D/\varepsilon_0\varepsilon$, the elec-

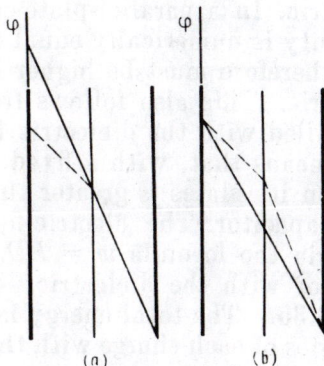

Fig. 3.33

tric field strength is lower in the half filled with the dielectric (where the potential gradient is smaller in absolute value), that is, part *1* (see the figure accompanying the ploblem). If removal of the dielectric does not alter the charge on the plates, the potential behaves in the same way as it did in part *2* prior to removal of dielectric and the total potential difference will increase (Figure (a) accompanying the answer). If removal of the dielectric does not alter the potential difference, the points representing the potentials on the plates (φ and 0) will remain unchanged, while the slope of the straight line will acquire a value intermediate between the one it had in the dielectric and in the air prior to removal of dielectric (Figure (b) accompanying the answer).

3.34. Since the lines of force of the electric displacement vector are continuous and the field in each part is uniform, with the lines of force being perpendicular to the vacuum-dielectric interface, the electric displacements is the same in both parts. The electric field strength, which is defined by the formula $E = D/\varepsilon_0\varepsilon$, is higher in the vacuum. The electric-field energy density is determined via the formula $w = ED/2$, which shows that this quantity is higher in the vacuum.

3.35. Since the potential difference between the plates of the two capacitors is the same and so is the distance between the plates, the electric field, which for a parallel-plate capacitor is $E = U/l$, is the same for both capacitors. According to its definition, the electric displacement $D = \varepsilon_0\varepsilon E$, is greater in the capacitor with the dielectric. In a parallel-plate capacitor, the surface charge density is numerically equal to the electric displacement and therefore must be higher in the capacitor with the dielectric. This also follows from the fact that the capacitor filled with the dielectric has a higher capacitance, which means that, with a fixed potential difference, the charge on its plates is greater than that on the plates of the air capacitor. The electric-field energy density, determined via the formula $w = ED/2$, is also higher in the capacitor with the dielectric.

3.36. The total energy is the sum of the interaction energies of each charge with the other charges in the system, or

$$W = \frac{Q^2}{4\pi\varepsilon_0\varepsilon r} - 2\,\frac{QQ_1}{4\pi\varepsilon_0\varepsilon r}\,.$$

By hypothesis, $W = 0$, whence $Q_1 = Q/2$.

3.37. The energy stored by a capacitor is determined by the electric-field energy density in the capacitor and the capacitor's volume: $W = wSl$. Since the energy density is

$$w = D^2/\varepsilon_0\varepsilon, \tag{3.37.1}$$

after the dielectric is removed, the energy of the capacitor will increase ε-fold. Since the charge on the capacitor remains unchanged, the value of the electric displacement vector remains unchanged, too. If prior to removal of the dielectric the distance between the plates was l_1 and after removal it was changed and became equal to l_2, the fact that the energy remained unchanged in the process can be expressed as follows:

$$\frac{D^2 S l_1}{\varepsilon_0 \varepsilon_1} = \frac{D^2 S l_2}{\varepsilon_0 \varepsilon_2} .$$

Hence the distance between the plates must be decreased ε-fold. Formula (3.37.1) shows that after the dielectric is removed (but prior to changing the distance between the plates) the capacitor increases its energy. This increase in energy is due to the work performed in removing the dielectric. The work is done against the forces of attraction of the free charges on the plates of the capacitor and the bound charges on the surface of the dielectric.

3.38. Since the capacitor voltage remains constant, the energy stored in the capacitor, $W = U^2 C/2$, decreases because when the dielectric is removed, the capacitance decreases ε-fold. If the entire system consisting of the DC source and the capacitor is considered, it can be seen that the charge flows from the capacitor to the source when the dielectric is being removed. A fraction of the energy stored in the capacitor is spent on heating the leads that connect the capacitor with the source of potential, while still another fraction goes into the source. Note that removing the dielectric from the capacitor requires performing mechanical work, which must be included in the general energy balance. It is expedient, for the sake of comparison, to consider the reverse process, the introduction of a dielectric into the capacitor. Since in this case the capacitance of the capacitor grows, the energy grows, too. This growth is provided by the energy stored in the source (a DC source), which supplies the capacitor with the necessary charge as the capacitance is increased.

3.39. The problem can be related to Problem 3.38. The answer can be obtained from the general formula for the energy stored in a charged capacitor: $W = Q^2/2C$. When the capacitor is submerged into liquid dielectric, its capacitance increases, with the result that the energy stored by the capacitor decreases, since the charge on it remains unchanged. Thus, if the liquid dielectric is "sucked" into the capacitor, the capacitor-dielectric system goes over to a state with a lower energy. This process continues until the decrease in energy is compensated for by the increase in the potential energy of the layer of dielectric between the plates in the gravitation field of the earth. It must also be noted that work is done against viscosity forces when the capacitor is drawn out or submerged into the dielectric. After the capacitor is submerged into the dielectric, its capacitance will increase, while the potential difference between the plates will drop. The electric field strength, which is the same in the parts with and without the dielectric, decreases too, while the electric displacement proves to be greater by a factor of ε in the part with the dielectric as compared to the value in the part without the dielectric.

3.40. The problem can be related to Problem 3.38. There we found that into the general energy balance one must include the energy flow through the current source, which uses a fraction of its energy to increase the energy stored in the capacitor when the capacitor is submerged into the dielectric. The liquid dielectric must be "sucked" into the capacitor, just as in the previous problem. The effect of the capacitor's field on the dielectric can also be taken into account by considering the polarization of the dielectric. As a result of this process, each volume element of the dielectric becomes a dipole and is pulled into the field at the edge of the capacitor. The strength of this field is higher than that in the dielectric at a certain distance from the plates.

3.41. When the cube is compressed in the transverse direction, it is stretched in the longitudinal direction, as a result of which the upper face becomes negatively charged and the lower face becomes positively charged.

3.42. Formally, such points are determined by the expression

$$\varepsilon = D/\varepsilon_0 E.$$

Obviously, at the point where $E = 0$ and $D \neq 0$ (point 0), the dielectric constant is formally equal to infinity, while at points where $D = 0$ and $E \neq 0$ it is zero (points 3 and 6). Of course, such values of ε are of a purely formal nature.

3.43. If l is the length of the plates of the capacitor in the system where the capacitor is at rest, in a system where the capacitor is moving with a velocity v this length is $l\sqrt{1 - v^2/c^2}$. Since the transverse dimensions of the plates do not change, the area ratio is also $1/\sqrt{1 - v^2/c^2}$. Since the charge on the capacitor remains unchanged, the surface charge density increases, with the result that $E/E_0 = 1/\sqrt{1 - v^2/c^2}$.

4. Direct Current

4.1. The two conductors, *1-3-5* and *2-4-6*, have different potentials, with the result that when key K is closed, a current will flow from *3* to *4*, while the currents passing through the resistors will flow from *1* to *3*, from *5* to *3*, from *4* to *6*, and from *4* to *2*. The closing of the key leads to an increase in the current flowing through the ammeter. If the resistances of the conductors *1-3-5*, *2-4-6*, and *3-4* are extremely low, then the sections *1-2* and *5-6* of the resistors will be shorted for all practical purposes.

4.2. Prior to closing the key, the circuit consists of two resistors connected in parallel (the resistance of each resistor being $3R$). This means that the total resistance of the circuit is $1.5R$. After the key has been closed, the circuit consists of two sections connected in series, each of which has two resistors connected in parallel. The resistance R' of each section is given by the formula

$$\frac{1}{R'} = \frac{1}{R} + \frac{1}{2R}$$

and is equal to $2R/3$. The resistance of the entire circuit is $4R/3$. The current measured by the ammeter is higher than that measured prior to closing the key.

4.3. If R is the resistance of the whole potentiometer and R_V is the resistance of the voltmeter, the total resistance of section ab of the potentiometer is

$$R_{ab} = \frac{R_V (R/2)}{R_V + R/2} = \frac{R}{2(1 + R/2R_V)} < \frac{R}{2}.$$

The resistance of section bc is equal to $R/2$. The voltage applied to the potentiometer will not be distributed evenly. Since the resistance of ab is less than that of bc, the voltage applied to the first section is lower than that applied to the second. The higher the resistance of the voltmeter, the closer the readings of the voltmeter are to one-half of the applied voltage.

4.4. Since the voltage applied to the "black box" is supplied by a DC source, it is natural to assume that there are only resistances inside the "box". The simplest way to lower voltage is to use a potentiometer (see the figure accompanying the answer). However, there is not a single circuit employing only resistances that can raise voltage. As the figure accompanying the answer demonstrates, from the voltage applied to terminals *1* and *2* one can always "take" a certain fraction, e.g. 127 V, while the 127 V applied to terminals *3* and *4* will yield the same 127 V

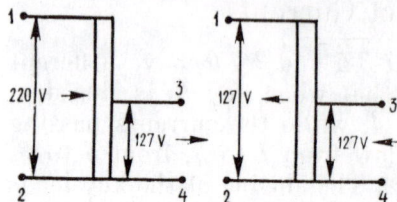

Fig. 4.4

on terminals *1* and *2*. The remark (made in the problem) that concerns the role of the measuring device is important since a voltmeter, which always has a finite resistance, redistributes the resistances in the circuit and, hence, changes the voltages (see Problem 4.3).

4.5. Let us assume, for the sake of simplicity, that the resistances of the two potentiometers are the same. When the sliding contact of each potentiometer is in the middle, the total resistance of the circuit is $R_0/2$, where R_0 is the resistance of each potentiometer. If the sliding contact of the second potentiometer is in the extreme (left or right) position, we have two resistances, R_0 and $R_0/2$, connected in parallel (assuming that the wires have no resistance), so that the total resistance is $R = R_0/3$. The reading of the ammeter proves to be greater than when the sliding contact of the second potentiometer was in the middle position by a factor of 1.5. Thus, when the sliding contact of the second potentiometer is moved from one extreme position to the other, the readings of the ammeter pass through a minimum.

4.6. If x is the resistance of the potentiometer between point a and the sliding contact, the total resistance between a and the sliding contact is $rx/(r + x)$, while the resistance of the entire circuit is $R - x + rx/(r + x)$. The current supplied by the DC source is

$$I = \frac{U_0}{R - x + rx/(r + x)} .$$

The potential difference between the sliding contact and point a is

$$U = \frac{U_0 rx}{(R-x)(r+x)+rx} = \frac{U_0 rx}{Rx-x^2+Rr}.$$

The current passing through the ammeter is

$$I_a = \frac{U_0 r}{Rx-x^2+Rr}. \qquad (4.6.1)$$

To find the extremum, we take the derivative

$$\frac{dI_a}{dx} = U_0 r \left[\frac{-R+2x}{(Rx-x^2+rx)^2} \right]. \qquad (4.6.2)$$

Nullifying (4.6.2) yields

$$x = R/2. \qquad (4.6.3)$$

If we substitute (4.6.3) into (4.6.1) we find the minimal current:

$$I_{min} = \frac{U_0 r}{R(r+R/4)}.$$

Thus, as the sliding contact is moved, the current through the ammeter passes through a minimum, and the

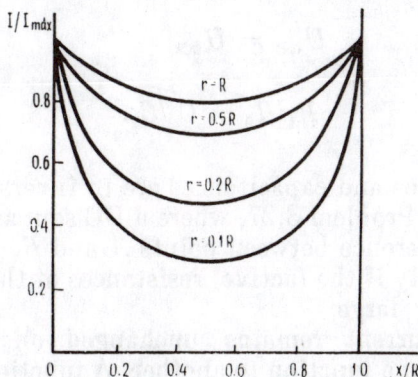

Fig. 4.6

smaller the r the deeper the minimum. At $x=0$ and $x=R$, a current of $I_{max} = U_0/R$ passes through the ammeter. The ratio of I_{min} to I_{max} is equal to $(r/R)(1+r/R)^{-1}$. The I_a/I_{max} vs. x/R curves for several values of r/R are shown in the figure accompanying the answer.

4.7. The exact value of $\mathcal{E}_x$ can be determined if one measures exactly the potential difference between points a and b provided that the current passing through the source in question is zero. This can be achieved by selecting a proper ratio of resistances between points a and b and points b and c using the resistance box. Knowing the resistance R between points a and b and the current I, measured by the ammeter, we find the sought emf:

$$\mathcal{E}_x = IR.$$

4.8. Since the currents in the resistors R_1 and R_2 are the same, we can write

$$U_{Aa}/R_1 = U_{aB}/R_2.$$

The charges on capacitors connected in series are the same, which means that

$$U_{Ab}C_1 = U_{bB}C_2.$$

Since

$$U_{Aa}/U_{aB} = R_1/R_2, \quad U_{Ab}/U_{bB} = C_2/C_1, \quad U_{Aa} = U_{Ab},$$

$$U_{aB} = U_{bB},$$

we have

$$R_1/R_2 = C_2/C_1.$$

The resistances and capacitances are in inverse ratio.

Just as in Problem 3.27, where a DC source generates a potential difference between points A and B, the solution holds true only if the (active) resistances of the capacitors are infinitely large.

4.9. The current remains unchanged on the entire section from one junction to another. A junction is a point in a circuit where more than three conductors meet. There are seven such sections in the figure accompanying the problem. If there are n junctions in a circuit, then Kirchhoff's first law yields $n - 1$ independent equations. There are four junctions in the circuit in question (*1, 4, 5* and *7*). Thus, to determine seven currents we are lacking four equations, which Kirchhoff's second law will yield. The simplest way to employ Kirchhoff's sec-

216

ond law is to use loops that do not overlap, namely, *1-2-3-4-1*, *1-4-5-1*, *1-5-7-1*, and *5-6-7-5*.

4.10. There are eight junctions in the circuit. Since Kirchhoff's first law yields only $n - 1$ independent equations for n junctions, we have seven such equations. If the circuit is transformed onto a plane (see the figure accompanying the answer), there are five nonoverlapping loops, while the loop *1-4-8-5-1* overlaps all other loops and therefore can be obtained from these.

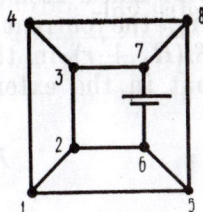

Fig. 4.10

4.11. The current flowing in the circuit is $I = \mathscr{E}/(R + r)$. The power output in the external circuit is

$$P = I^2 R = \mathscr{E}^2 \, \frac{R}{(R+r)^2} \, .$$

The maximal power output can be found from the condition $dP/dR = 0$, or

$$\frac{dP}{dR} = \mathscr{E}^2 \, \frac{(R+r)^2 - 2(R+r)R}{(R+r)^4} = 0,$$

whence $R = r$. The fact that the resistances are equal means that the power outputs must be equal, too:

$$I^2 r = I^2 R.$$

Hence, the efficiency is equal to 0.5.

4.12. The current is maximal when the circuit is shorted, or when the external resistance is zero:

$$I_{\mathrm{m}} = \mathscr{E}/r.$$

Thus, in both cases the ratio of the emf to the internal resistance is the same.

Maximal useful power output (the power output of the external resistance) is achieved when the exernal resistance is made equal to the internal resistance (see the answer to Problem 4.11), that is, when the current is one-half the maximal current. This power output is

$$P = \frac{\mathscr{E}^2}{4R} = \frac{\mathscr{E}^2}{4r} \, .$$

Since the ratio $\mathscr{E}/r$ is the same in both cases, a double useful power output is achieved at a double electromotive

217

force (for equal currents). Note that the internal resistance of the DC source must also be doubled if we want the ratio to remain unchanged.

4.13. If in one position of the sliding contact the rheostat has a resistance R_1 and in the other, a resistance R_2, the current is $\mathscr{E}/(R_1 + r)$ in the first case and $\mathscr{E}/(R_2 + r)$ in the other. Correspondingly, the power output in the external circuit (the same in both cases) is

$$P = \frac{\mathscr{E}^2 R_1}{(R_1 + r)^2} = \frac{\mathscr{E}^2 R_2}{(R_2 + r)^2}.$$

Dividing this expression by $\mathscr{E}^2$ and solving for r, we find that

$$r = \sqrt{R_1 R_2}.$$

By hypothesis, in one case $R_1 = xR$ and in the other, $R_2 = (1 - x) R$. Whence

$$r = R \sqrt{x(1 - x)}.$$

4.14. The likely circuit, apparently, consists of a combination of cells connected in parallel and in series. There are two possibilities here: several parallel groups of cells connected in series or several in-series groups of cells connected in parallel. First, it can be shown that the two variants are equivalent. Indeed, in the first variant, the potentials at the points a_1, a_2, a_3, etc. coincide, i.e.

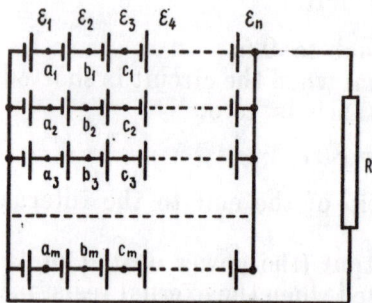

Fig. 4.14

$$U_{a_1} = U_{a_2} = U_{a_3} = ...;$$

the same is true of the potentials at the points b_1, b_2, b_3, etc., i.e.

$$U_{b_1} = U_{b_2} = U_{b_3} = ...$$

This line of reasoning can be continued. The respective points can be interconnected, and the entire circuit will be transformed into the second variant. Suppose the overall number of cells is N. We connect these cells in such a manner that groups of n cells that form $m = N/n$ parallel groups are connected in series. In this case the current in the external circuit is

$$I = \frac{\mathscr{E}n}{R + rn/m} = \frac{\mathscr{E}Nn}{RN + rn^2}.$$

The power output in the external circuit is

$$P = I^2 R = (\mathscr{E}N)^2 R \frac{n^2}{(RN + rn^2)^2}.$$

To find the maximal value, we nullify the derivative of P with respect to n:

$$\frac{dP}{dn} = \frac{2\mathscr{E}^2 N^2 n}{(RN + rn^2)^2}(RN - rn^2) = 0.$$

Whence

$$n = \sqrt{RN/r}. \tag{4.14.1}$$

But this does not solve the problem completely. The number n should be one of the cofactors of N. To find a practical value of n, we must compare the power outputs for two values of n that are closest to the one given by (4.14.1), that is, one must be smaller than the calculated value and the other must be greater, and yet the two must be cofactors of N. Here is an example. Suppose $N = 400$, $R = 16\ \Omega$ and $r = 9\ \Omega$. The calculated value is

$$n = \sqrt{\frac{400 \times 16}{9}} = 26.7.$$

The closest cofactors of N are 25 and 40. In this example the greater power output is at $n = 25$. Thus, the circuit consists of 16 parallel groups of 25 cells connected in series in each group.

4.15. Since the displacement current is defined as

$$I_{dis} = S \frac{dD}{dt},$$

after performing certain manipulations we can write

$$I_{dis} = \frac{\varepsilon_0 \varepsilon}{l} \frac{dU}{dt} = \frac{dQ}{dt},$$

where Q is the charge on the capacitor. Thus, the displacement current may be made constant over a definite time interval if the capacitor is charged (or discharged) by a direct current. For this in the circuit of the capacitor being charged we must have a device that restricts the current flowing through it within broad voltage limits (Figure (a)). A diode operating in the saturation mode may serve as such a device. For the case of a thermionic valve (or diode) the appropriate circuit is shown in Figure (b), while for the case of a semiconductor diode the cir-

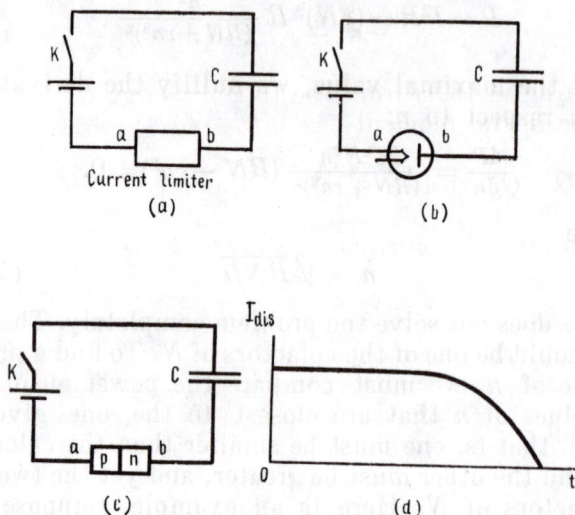

Current limiter

(a)

(b)

(c)

(d)

Fig. 4.15

cuit is shown in Figure (c). The diode is introduced into the circuit in the cut-off direction, and the voltage across the diode is

$$U_d = \mathscr{E}_0 - U_C.$$

As long as U_d remains within the saturation region, the current through the diode (and, hence, the charging current) remains constant. The displacement current remains constant in the process. After a certain time interval has elapsed (the lower the charging current the longer the interval), the charging current rapidly falls off to zero. The time dependence of the displacement current is illustrated schematically in Figure (d).

4.16. At each moment of time the capacitor voltage is equal to the potential drop across the resistor;

$$U = IR.$$

Bearing in mind that $U = Q/C$ and $I = -dQ/dt$ (the minus sign shows that the capacitor's charge decreases), we get

$$\frac{Q}{C} = -R\,\frac{dQ}{dt},$$

or

$$\frac{dQ}{Q} = -\frac{1}{RC}\,dt. \qquad (4.16.1)$$

Integrating (4.16.1) from the initial charge Q_0 to Q and from the initial moment $t = 0$ to time t, we get

$$Q = Q_0 \exp\left(-\frac{t}{RC}\right) = U_0 C \exp\left(-\frac{t}{RC}\right).$$

Accordingly, the current varies with time as follows:

$$I = I_0 \exp\left(-\frac{t}{RC}\right), \qquad (4.16.2)$$

with $I_0 = U_0/R$. Taking logs, we can write (4.16.2) as follows

$$\ln I = \ln I_0 - \left(\frac{1}{RC}\right) t.$$

Thus, the time dependence of $\ln I$ is represented by a straight line with a negative slope, whose absolute value is $1/RC$. The resist-
ance R determines the current at the first moment of discharge and the initial capacitor voltage, which is equal to the emf of the source. The value of R determined in this manner and the slope of the straight line fix the value of C.

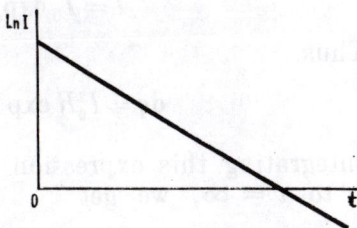

Fig. 4.16

4.17. As shown in the answer to Problem 4.16, the discharge current varies with time as

$$I = I_0 \exp\left(-\frac{t}{RC}\right), \quad \text{or} \quad \ln I = \ln I_0 - \left(\frac{1}{RC}\right) t.$$

Initially, i.e. at $t = 0$, both currents are the same (see the figure accompanying the problem). For a fixed capacitance C this is possible if the other two parameters, U

and R, change simultaneously. Since by hypothesis only one parameter changes, we conclude that the capacitance C varies. The fact that the slope of the straight line representing the $\ln I$ vs. t dependence decreases means that the capacitance C increases.

4.18. The time variation of the current proceeds as follows:

$$\ln I = \ln I_0 - \left(\frac{1}{RC}\right) t.$$

The fact that the two straight lines, 1 and 2, are parallel indicates that the product RC must be constant. Since by hypothesis the discharge processes differ only in the value of one parameter, both R and C remain constant. What is different is the initial capacitor voltage, and since for straight line 1 the initial current is higher than for straight line 2, so is the initial capacitor voltage.

4.19. The current flowing through the resistor with resistance R will generate during a time interval dt the following amount of heat:

$$dq = I^2 R dt.$$

The time variation of the discharge current of the capacitor is

$$I = I_0 \exp\left(-\frac{t}{RC}\right).$$

Thus,

$$dq = I_0^2 R \exp\left(-\frac{2t}{RC}\right) dt.$$

Integrating this expression with respect to t from $t = 0$ to $t = \infty$, we get

$$q = I_0^2 RC/2. \qquad (4.19.1)$$

At the first moment the discharge current is

$$I_0 = U_0/R. \qquad (4.19.2)$$

Substituting (4.19.2) into (4.19.1), we obtain the initial energy stored by the charged capacitor:

$$q = U_0^2 C/2.$$

4.20. According to Kirchhoff's second law, at each moment of time the emf of the DC source is equal to the sum of the potential drop across the resistor and the capacitor voltage:

$$\mathcal{E} = IR + U.$$

Bearing in mind that $I = dQ/dt$ and $U = Q/C$, we get

$$\mathcal{E} = R\,\frac{dQ}{dt} + \frac{Q}{C}, \quad \text{or} \quad \frac{dQ}{Q - \mathcal{E}C} = -\frac{1}{RC}\,dt.$$

Integration from $Q = 0$ to Q and from $t = 0$ to t and appropriate transformations yield

$$Q = \mathcal{E}C\left[1 - \exp\left(-\frac{t}{RC}\right)\right],$$

whence

$$I = \frac{\mathcal{E}}{R}\exp\left(-\frac{t}{RC}\right) = I_0\exp\left(-\frac{T}{RC}\right).$$

The amount of heat generated by the current in the resistor R in the course of dt is

$$dq = I^2R\,dt = I_0^2R\exp\left(-\frac{2t}{RC}\right).$$

Integration from $t = 0$ to $t = \infty$ yields

$$q = I_0^2R^2C/2 = \mathcal{E}^2C/2.$$

The same amount of energy is stored by the capacitor when the latter is charged to a voltage equal to the source's emf. The total energy used up by the source,

$$\int\limits_0^\infty \mathcal{E}I\,dt = \mathcal{E}I_0\int\limits_0^\infty e^{-t/(RC)}\,dt,$$

is equal to $\mathcal{E}^2C$, which is the sum of two equal quantities $\mathcal{E}^2C/2$.

4.21. The energy stored by a charged capacitor can be written in the form

$$W_0 = Q^2/2C.$$

After the second (uncharged) capacitor is connected to the first, the total charge does not change while the capacitance doubles. Thus, the total energy stored by this system becomes

$$W = Q^2/4C,$$

which is one-half of the initial energy. So where did the other half go to? As the charge is redistributed between the two capacitors, a current flows through the conductors connecting them and generates heat. In addition, there is always a magnetic field around a conductor with current, and this magnetic field carries energy, just as an electric field does. If the resistance of the conductors is low (zero in the case of superconductors), the difference between the initial and the final energy will go to the magnetic field. Eventually the second capacitor will become fully charged while the first capacitor will become completely discharged and the current will cease. Then the second capacitor will begin to discharge, and charge will flow to the first capacitor. This process will continue, that is, there will appear electromagnetic oscillations in which the energy will alternate between that of the electric field and that of the magnetic.

4.22. For an electron that is inside the disk at a distance r from the axis to move along a circle, there should be a force pulling it to the axis. According to Newton's second law,

$$F = m\omega^2 r.$$

This force is generated by a radial electric field caused by the redistribution of the electrons in the disk and is such that the force acting on the electron is

$$F = eE = m\omega^2 r.$$

If we substitute $-d\varphi/dt$ for E and integrate from φ_1 to φ_2 and from 0 to R, where R is the radius of the disk, we get

$$\int_{\varphi_1}^{\varphi_2} \frac{d\varphi}{dt} = -\frac{m\omega^2}{e} \int_0^R r\,dr.$$

As a result, we get the potential difference between the center of the disk and the edge:

$$U = \varphi_1 - \varphi_2 = \frac{m\omega^2 R^2}{2e} = \frac{mv^2}{2e}, \qquad (4.22.1)$$

where v is the linear velocity of points at the edge of the disk. Theoretically formula (4.22.1) can be used to determine the electron's charge-to-mass ratio. But actually this constitutes a problem, as shown by an estimate of the potential difference between the axis and the edge. The electron charge is 1.6×10^{-19} C and the electron mass is 9.1×10^{-31} kg. We set the electron linear velocity on the edge at 300 m/s. The potential difference then proves to be less than 10^{-9} V. It is extremely difficult to measure such a quantity in such a rotating system.

4.23. Moving together with the cylinder, the electrons in the wire have a momentum mv each. When the cylinder is braked, the electrons continue to move, but the generated potential difference creates a braking electric field of strength E. The force acting on every electron in the wire is

$$F = eU/l,$$

with U the instantaneous potential difference. According to Newton's second law,

$$\Delta mv = \frac{e}{l} \int_0^t U\, dt,$$

where Δmv is the momentum lost by an electron during the entire braking time, which quantity is equal to the initial momentum mv. The charge-to-mass ratio for the electron is then

$$\frac{e}{m} = \frac{vl}{\int_0^\infty U\, dt}.$$

The integral in the denominator can be evaluated by calculating the area under the voltage oscillogram.

4.24. The heating of the conductor will result in the electron diffusing into the neighborhood of section ab, with the potential of the conductor somewhat increasing. The current flowing in the conductor will have to overcome a potential barrier at point a. This requires additional energy, which will be taken from the metal. On the other hand, when passing through the conductor at point b, the current goes to a region with a lower potential, and in this place energy will be released to the metal. As a result, the point where the temperature is at a maximum will shift in the direction of current flow.

4.25. Prior to cooling, the resistance of the wire was the same over the entire length of the wire (precisely, the resistivity was the same at all points of the wire). When the fan is switched on, the resistance of the section that is being cooled will lower. This leads to a redistribution of the potential between the cooled and uncooled sections, with the greater voltage applied to the latter section, as a result of which its temperature increases. This phenomenon is enhanced by the fact that the resistance of the uncooled section somewhat grows with temperature, which leads to a still greater inhomogeneity in the distribution of the potential in both sections.

4.26. The resistances of bulbs with the same rated voltage are in inverse proportion to the rated wattages. Hence, the resistance of the bulb with the lower wattage is six times the resistance of the bulb with the higher wattage. When the bulbs are connected in series with the DC source, the current is the same and six-sevenths of the total voltage of 220 V, or 189 V, is applied to the first (25 W) bulb and one-seventh, to the second (150 W) bulb. Actually the difference is still greater because the resistance of the first bulb will increase due to overheating, while that of the second will decrease. Hence, the 25-W bulb must burn out.

4.27. An increase in voltage will lead to an increase in the currents passing through the conductor and semiconductor, and this will lead to an increase in temperature of both. As a result the resistance of the conductor will increase and that of the semiconductor will decrease. Hence, the current through the semiconductor will increase greater than in proportion to the voltage, while the current through the conductor will increase lesser than in proportion to the applied voltage, with the result that the ammeter in the semiconductor circuit will register a higher current than the ammeter in the conductor circuit.

4.28. Prior to an increase in voltage, the resistances of the semiconductor and the conductor were equal. When the voltage is increased, the current in the circuit increases, too, and so does the temperatures of the semiconductor and conductor. This leads to a drop in the resistance of the semiconductor and an increase in the resistance of the conductor. The voltage between the semiconductor and conductor will redistribute in such a manner that the voltmeter connected to the conductor will register a higher voltage than the voltmeter connected to the semiconductor.

4.29. The electrons leaving the filament, or cathode, create a negative space charge whose field does not let all the emitted electrons into the region. According to the Child-Langmuir theory developed for parallel plane electrodes on the assumption that the initial velocity of the electrons is zero, the current density between the electrodes is

$$j = \frac{4\sqrt{2}\,\varepsilon_0}{9} \sqrt{\frac{e}{m}} \frac{U^{3/2}}{d^2}$$

(the three-halves power law). Here e and m are the electron charge and electron mass, U is the voltage drop across the electrodes, and d is the distance between the electrodes. On the current-voltage characteristic, the initial segment of the curve agrees with the three-halves power law. Then, as the electron cloud is dissipated, the current gradually reaches a plateau and saturation sets in, with the saturation current being the total flux of electrons that the cathode can deliver at a given temperature. The temperature dependence of the current density is given by the Richardson-Dushman equation

$$j_{\text{sat}} = A' T^2 \exp\left(-\frac{P}{kT}\right).$$

The quantity P in the numerator of the exponent is the so-called work function, or the work that an electron must do to leave the metal. The other quantities in the equation are as follows: T the thermodynamic temperature, k the Boltzmann constant, $A' = 6.02 \times 10^5$ A/m^2·K^2 is a constant that is a combination of universal constants. The difference in the curves in the figure accompanying the problem lies in the temperature of the cathode, which is higher for curve 2.

4.3 . When thermoelectric current flows from the cathode to the anode, the electrons leaving the cathode carry away an energy required for overcoming the potential barrier that exists at the metal-vacuum interface (the work function of the electrons), with the result that the cathode cools off. To maintain a constant cathode temperature, the filament current must be increased.

4.31. When the potential difference between the electrodes is nil, the concentration of positive and negative ions (cations and anions) is the same in practically the entire volume. When an external voltage is applied, a current generated by the motions of cations to the cathode and anions to the anode begins to flow in the electrolyte. As a result, the regions near the electrodes prove to be depleted of ions whose sign is that of the electrode. Cations leave the anode and anions leave the cathode. For this reason, near the anode an excess of negative charge is formed, while an excess of positive charge is formed in the region near the cathode. All this leads to a distortion in the electric field. The enhanced field near the electrodes imparts an enhanced velocity to the ions. This ensures the flow of current under lower charge carrier concentrations.

4.32. The sign of the volume charge is determined by the direction of convexity of the $\dot{U}$ vs. x curve.* The volume charge is positive where the curve is convex upward and negative where the curve is convex downward, while the volume charge is nil where the U vs. x dependence is represented by a straight line. Hence, the entire region between the cathode and the anode is divided, within the first approximation (i.e. ignoring certain details), into the cathode space (from point 0 to point 1 in the figure accompanying the problem) with a surplus positive charge, the Faraday dark space (from point 1 to point 2) with a negative charge, and the region of the "positive column" (from point 2 to point 3), which constitutes a plasma with practically equal concentrations of electrons and positive ions and, hence, with a net charge that is practically nil.

* See Problems 3.28 and 3.29.

4.33. The conduction-current density is given by the formula

$$j = e \sum_k n_k u_k Z_k, \qquad (4.33.1)$$

where e is the magnitude (without taking into account the sign) of the elementary charge (the electron charge), n_k the concentration of the given type of charge carriers, u_k the average directional velocity of the carriers, and Z_k the charge number, or valence, of the carriers. For an electron $Z = -1$, while for a positive doubly charged ion (say, He^{++}) $Z = +2$. Electron velocities exceed ion velocities by a factor of 10 or even 100, with the result that even at equal concentrations the electron current is much stronger than the ion current. Since in an electric field electrons and ions move in opposite directions, we can assume that the electron velocity is negative if the ion velocity is set positive. Since the number Z for electrons is negative, the signs of the products in (4.33.1) coincide, with the result that the ammeter in the gas discharge gap circuit will register the total current of electrons and ions.

4.34. As the particle moves from the anode to the Faraday cylinder, the field in the region between A and F constantly changes. When the particle leaves the anode (through the aperture) and is moving toward the Faraday

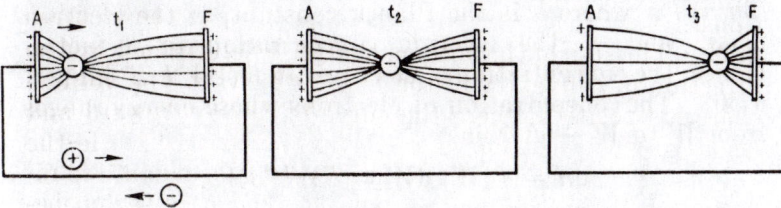

Fig. 4.34

cylinder, it induces positive charges on these electrodes, and the magnitude of these charges constantly changes. The density of these charges on the anode decreases while that on the Faraday cylinder increases (the variation in the distribution of electric charge for three moments in time is shown in the figure accompanying the answer). For this reason, in the region of space between A and F there appears a continuous displacement current, which means that an exact replica of this current appears in the circuit. The current in the circuit can be graphically represented as a consequence of the fact that in approaching the Faraday cylinder the particle repels, so to say, the electrons which, in effect, move toward the anode through the measuring device G. Thus, the current in the circuit exists during the entire time of motion of the particle between the anode and the Faraday cylinder, as shown in Figure (e) accompanying the problem.

4.35. If two metals are brought into contact, the limiting energies of the electrons will establish themselves at a common level (the common Fermi level; see the figure accompanying the answer). The difference between the height of a potential barrier and the Fermi level determines the external work function $e\varphi$. The difference between the two work functions (for the two barriers) is equal to the external contact potential difference. To transfer an electron from the surface of metal *2* to the surface of metal *1* requires performing an amount of work equal to

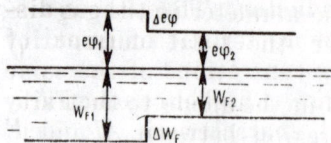

$e\Delta\varphi$. The distance between the levels of minimal electron energy in the metals determines the internal contact potential difference ΔW_F. According to the quantum theory of metals, the Fermi level at 0 K is pinned at $(h^2/2m) \times$

Fig. 4.35

$(3n/\pi)^{2/3}$, where h is the Planck constant, m the electron mass, and n the electron concentration in a metal. Hence, the concentration of electrons in metal *1* is higher.

4.36. The concentration of electrons whose energy ranges from W to $W + \Delta W$ is

$$dn = f(W)\,dW = CW^{1/2}dW,$$

in accordance with Eq. (4.36.1). Integrating this expression from zero to the limiting energy, we obtain the concentration of electrons in the entire energy range:

$$n = C \int_0^{W_F} W^{1/2}\,dW = \frac{2}{3}CW_F^{3/2}.$$

Hence, $W_F \propto n^{2/3}$. As is proved in the quantum theory of metals, W_F is given by the following formula (with due regard for universal constants):

$$W_F = \frac{h^2}{2m}\left(\frac{3n}{\pi}\right)^{2/3}.$$

4.37. The electrical conductivity (specific conductance) of a semiconductor depends on temperature according to the following law:

$$\sigma = \sigma_0 \exp\left(-\frac{W}{kT}\right), \quad \text{or} \quad \ln\sigma = \ln\sigma_0 - \frac{W}{kT},$$

where W is the width of the forbidden band. This law implies that the wider the forbidden band, the steeper the straight line representing the $\ln\sigma$ vs. T^{-1} dependence. Hence, semiconductor *1* has a wider forbidden band.

4.38. Since the upper sections of the curves for the two semiconductors coincide and the slopes of the lower sec-

230

tions are smaller than those of the upper, the intrinsic conductivities are the same. Also, since the lower sections of the curves slope in the same manner (i.e. the slopes are equal), the width of the forbidden band for the impurities is the same for both semiconductors. Thus, these diagrams can be interpreted as a characteristic of impurity semiconductors with different concentrations of the same impurities. For a fixed temperature, a higher conductivity corresponds to a higher located characteristic, *1*, and, hence, a higher concentration of impurities.

4.39. The diffusion of electrons from an n-type semiconductor into a p-type one and of holes from the p-type semiconductor to the n-type one leads to the appearance of a positive potential on the n-type semiconductor (the left branch of curve *0*) and a negative potential on the p-type semiconductor (the right branch of the same curve). If we now apply a positive potential to the n-type semiconductor and a negative potential to the p-type, the potential difference between the two semiconductors will increase (curve *1*), whence the boundary between the two semiconductors is depleted of charge carries as a result of electrons being drained to the n-type semiconductor and holes, to the p-type. Such a direction of the potential difference is the cut-off one. When the external voltage is applied in the reverse direction (curve *2*), the potential difference lowers, and it proves easier for the electrons to move to the p-type semiconductor and the holes, to the n-type. This direction is the conducting one.

4.40. Every semiconductor possesses intrinsic conduction in addition to extrinsic (or impurity) conduction. Intrinsic conduction is caused by the transfer of electrons from the valence band to the conduction band and by simultaneous formation of holes in the valence band. Intrinsic conduction is of a mixed nature for this reason, and because of this the n-type semiconductor carries a small number of holes while the p-type semiconductor carries a small number of electrons. When a voltage is applied in the cut-off direction, these charge carriers constitute the so-called reverse current. As the temperature of a semiconductor diode is increased, the electron and hole concentrations grow, as a result of which conductivity in the cut-off direction grows, too. The reverse current reaches a plateau when practically all the "alien" charge carriers (holes in the n-type semiconductor and electrons in

231

the p-type) participate. Usually this current is several orders of magnitude less than the direct current, with the plateau reached at relatively high voltages. The direct current grows with voltage very rapidly, since as the voltage is increased, it becomes easier for the charge carriers to pass through the junction.

4.41. Let us suppose that there is a small deviation from the state with $\sigma = 1$. If σ drops at point a, the number of electrons impinging on the surface is smaller than the number of electrons leaving the surface, with the surface acquiring a negative potential, which brings down σ still further. And this leads to a further increase in the negative potential. The process continues until the current of primary electrons becomes totally cut off. If at the same point the value of σ increases somewhat, the surface acquires a positive potential, the velocity of the electrons increases, and the current continues to grow, which leads to an increase in σ, up to the maximum on the curve, and then to point b, where $\sigma = 1$, just as at point a. Similar reasoning leads us to the conclusion that small variations in σ at point b change the potential of the surface in such a way that σ returns to its initial value $\sigma = 1$. Thus, point a corresponds to an unstable state, while point b corresponds to a stable state. For a small deviation from the state of equilibrium corresponding to point a, the surface acquires a potential that either completely cuts off the current of the primary electrons or corresponds to that at point b.

4.42. The reflected electrons retain practically all their initial energy and, hence, correspond to curve *2* in the figure accompanying the problem. Secondary electrons, on the other hand, are freed from the solid bombarded with the primary electrons at the expense of the energy of the primary electrons, and this energy is distributed between the emitted electrons. The energy of the latter is, as a rule, considerably less than that of the primary electrons. Moreover, while all the reflected electrons have velocities that are concentrated within a narrow interval and have energies close to those of the primary electrons, the secondary electrons form a broad spectrum of velocities. The "true" secondary electrons are represented by curve *1* in the figure accompanying the problem.

5. Electromagnetism

5.1. If we use the right-hand screw rule, we will find that both in region I and in region III the directions of the magnetic induction vectors coincide and the resultant induction may vanish only at infinity. The same rule shows that only in region II can the magnetic induction vectors point in opposite directions (i.e. the induction created by the two currents), with the resultant induction vanishing somewhere inside II. If a is the distance separating the conductors, then the distance x from a conductor carrying the current I_1 to the point where the induction is zero can be found from the equation

$$\frac{\mu_0 \mu I_1}{2\pi x} - \frac{\mu_0 \mu I_2}{2\pi (a-x)} = 0.$$

Hence,

$$x = \frac{I_1}{I_1 + I_2}\, a.$$

5.2. If we use the right-hand screw rule, we will establish that the magnetic induction can vanish only in sectors I and III. If y is the distance from a certain point on the conductor carrying the current I_1 to the point where the magnetic induction is zero, and x is the distance from this point to the conductor carrying the current I_2, then

$$\frac{\mu_0 \mu I_1}{2\pi y} = \frac{\mu_0 \mu I_2}{2\pi x}.$$

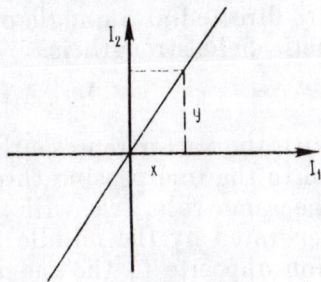

Fig. 5.2

Hence, the locus of points where the magnetic induction is zero is the straight line that passes through the point of intersection of the conductors and whose equation is $y = (I_1/I_2)\, x$.

5.3. The magnetic inductions generated by a straight conductor with a current and a circular conductor are, respectively,

$$B = \frac{\mu_0 \mu I}{2\pi r} \quad \text{and} \quad B = \frac{\mu_0 \mu I}{2r}.$$

233

In the case corresponding to Figure (b), the directions of the two induction vectors coincide, while in the case corresponding to Figure (c) they are opposite. Thus,

$$B_a = \frac{\mu_0 \mu I}{2r}, \quad B_b = \frac{\mu_0 \mu I}{2r}\left(1 + \frac{1}{\pi}\right),$$

$$B_c = \frac{\mu_0 \mu I}{2r}\left(1 - \frac{1}{\pi}\right),$$

whence

$$B_b = \frac{\pi + 1}{\pi} B_a = 1.32\,B_a, \quad B_c = \frac{\pi - 1}{\pi} B_a = 0.68\,B_b.$$

5.4. If the distance from the middle conductor to each of the other two conductors and to the point where we wish to determine the field is a, the magnetic field generated by each outer conductor at this point is

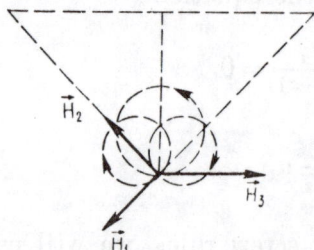

Fig. 5.4

$$H_1 = \frac{I}{2\pi a \sqrt{2}}.$$

Using the right-hand screw rule, we find that the vectors of the magnetic fields generated by the outer conductors are directed at an angle of 90°, so that the resultant magnetic field strength is

$$H_{1,2} = \sqrt{2}\,H_1 = \frac{I}{2\pi a},$$

with the vector representing this resultant directed parallel to the line passing through the conductors. Employing the same rule, we will find that the magnetic field H_3 generated by the middle conductor points in the direction opposite to the one of the resultant $H_{1,2}$, with $H_3 = I/2\pi a$, that is, $|H_3| = |H_{1,2}|$. Thus, the resultant of all three fields is zero.

5.5. A magnetic induction vector is always directed along a tangent to a line of force (for each of the four conductors the line of force is a circle in the plane of the drawing). As the figure accompanying the answer shows, the magnetic inductions generated by currents I_1 and I_4 are directed along the diagonal of the square from the conductor carrying I_2 to the conductor carrying I_3. Reasoning along the same line, we conclude that the magnetic inductions generated by currents I_2 and I_3 are directed along the diagonal of the square from the conductor car-* rying I_4 to the conductor carrying I_1. The resultant induction of the magnetic field generated by all four currents, or the geometric sum of the magnetic induction vectors of the four currents, lies in the plane of the drawing and points from right to left.

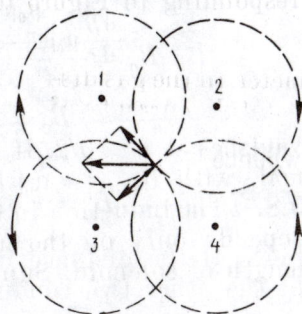

Fig. 5.5

5.6. The presence of a maximum in the middle between the conductors suggests that the currents in the conductors are flowing in opposite directions and that, the currents are equal in magnitude. Allowing for the direction of the induction vector at point M and employing the right-hand screw rule (see the figure accompanying the answer), we conclude that in the

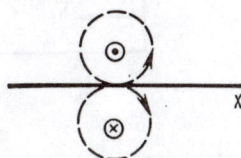

Fig. 5.6

upper conductor the current is directed toward the reader and in the lower, away from the reader.

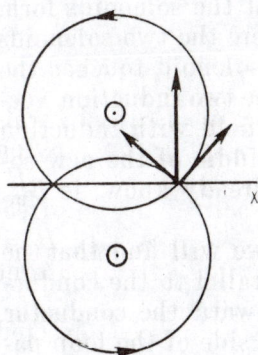

Fig. 5.7

5.7. At the point that lies in the middle between the conductors the induction is zero, which means that both currents flow in the same direction. Employing the right-hand screw rule, we can determine the direction of the magnetic induction vector in the region to the right of the conductors for both possible directions of current. As the figure accompanying the problem shows, the induction vector to the right of the conductors is directed upward. Hence, the currents are flowing toward the reader (see the figure accompanying the answer). At a distance x from point M the induction is

$$B = \frac{\mu_0 \mu I x}{\pi \left[x^2 + \left(\frac{a}{2} \right)^2 \right]}.$$

The derivative

$$\frac{dB}{dx} = \frac{\mu_0 \mu I \left[x^2 + \left(\dfrac{a}{2} \right)^2 - 2x^2 \right]}{\pi \left[x^2 + \left(\dfrac{a}{2} \right)^2 \right]^2}$$

vanishes at $x = a/2$. It is at this distance that B is maximal, with $B_{max} = \mu_0 \mu I / \pi a$.

5.8.   The induction in the middle of a very long solenoid depends only on the number of ampere-turns per unit length of solenoid. Suppose that we have two very long,

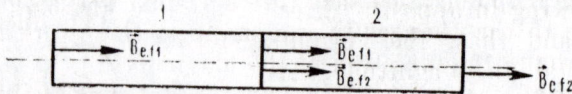

Fig. 5.8

similar solenoids with equal ampere-turns per unit length and that these solenoids are placed far apart. We denote the induction in the middle of a solenoid by B_m and that at an end face, by $B_{e.f}$. Let us bring these two solenoids together in such a manner that the directions of their magnetic inductions coincide and that the solenoids form a new long solenoid. At the point where the two solenoids touch (the right end face of the left solenoid touches the left end face of the right solenoid), the two induction vectors $B_{e.f}$ add up and form the total field with induction $2B_{e.f}$. But this point is simply the middle of the new solenoid, where the induction, as we already know, is B_m. Thus, $B_m = 2B_{e.f}$.

5.9.   Employing the left-hand rule, we will find that the force acting on the side of the loop parallel to the conductor and closest to it is directed toward the conductor while the force acting on the opposite side of the loop parallel to the conductor and farthest from it is directed away from the conductor. Since the first force is greater in magnitude, the loop moves toward the conductor. Employing the same rule once more, we will see that the force acting on the upper side of the loop is directed upward while that acting on the lower side of the loop is directed downward. Thus, the forces tend to pull the loop apart, that is, to increase the area subtended by the loop.

This will actually happen if the material of the loop is elastic.

The answer to the question can be obtained from a more general reasoning. The work done in the process of displacing a loop carrying a current in a magnetic field is $A = I\Delta\Psi$, where $\Delta\Psi$ is the increment of the magnetic flux coupled with the loop. The loop tends to move or change its form in such a manner that the magnetic flux coupled with it acquires the greatest possible value. The flux is assumed to be positive if inside the loop it coincides in direction with the flux created by the current in the loop. In allowing for the various changes in the flux coupled with the loop one must take into account the changes that are due to the changes in the shape of the loop. In the case at hand the direction of the magnetic flux created by the current in the straight conductor and that of the magnetic flux created by the current in the loop coincide, and since the induction of the field created by the current in the straight conductor increases as we move closer to the conductor, this will lead to a certain displacement of the loop. The fact that the square loop transforms into a circle as the loop's area increases also leads to an increase in both the outer and inner magnetic fluxes.

5.10. Both a torque and a force act on the loop. The direction of the torque is determined by the fact that the positive normal to the plane of the loop must point in the direction of the induction of the external field. The right-hand screw rule is used to determine the positive direction of this normal, which therefore coincides with the direction of the magnetic field of the loop proper. In accord with the direction of the current in the loop, the positive normal points upward. In the external field the loop turns counterclockwise, with the magnetic field generated by the current flowing in the loop coinciding in direction with the external magnetic field. The direction of the force acting on the loop is determined by the nature of the inhomogeneity of the external field. Since a loop carrying a current and placed in an external magnetic field moves in such a manner that the magnetic flux coupled with it attains the maximal possible value (in the algebraic sense), when the directions of the external and the intrinsic flux coincide, the motion occurs in the direction of the field with the higher induction, which in the case at hand means from left to right.

5.11. A similar question has been considered in Problem 5.9. Although in this case no external magnetic flux is present, the contour may influence the magnitude of the flux coupled with it by changing its shape. Since the area and, hence, the flux through the contour are maximal when the contour is in the form of a circle, the magnetic forces acting on the contour tend to transform the contour in just this manner. We can arrive at the same conclusion by considering the interaction of two elements of the contour that are opposite to each other. The currents that flow in these elements tend to move the elements apart, since they flow in opposite directions. The collection of all such forces tends to stretch the contour.

5.12. The following force acts on a contour carrying a current and placed in a nonuniform magnetic field with

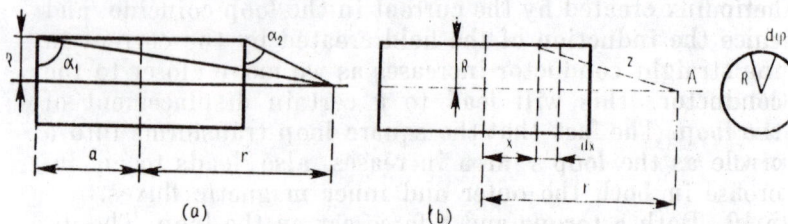

Fig. 5.12

the directions of the lines of force of this field coinciding with those of the field generated by the current in the contour: $F = p_m \dfrac{dB}{dr}$

In the case at hand, the force is determined by the values of the derivative dB/dr at different points of the field of the solenoid. The induction of the field of a solenoid of a finite length is given by the formula* (see Figure (a) accompanying the answer)

$$B = \frac{\mu_0 I N_0}{2} (\sin \alpha_1 - \sin \alpha_2). \qquad (5.12.1)$$

After simple transformations, the derivative dB/dr can be written as

$$\frac{dB}{dr} = \frac{\mu_0 I N_0}{2R} \left\{ \frac{1}{\left[1 + \left(\frac{r+a}{R}\right)\right]^{3/2}} - \frac{1}{\left[1 + \left(\frac{r-a}{R}\right)^2\right]^{3/2}} \right\}.$$

$$(5.12.2)$$

Formula (5.12.2) shows that dB/dr is nonpositive for $r > 0$. This means that the force acting on the contour is attractive (in Figure (a) this force points from right to left).

At $r = 0$ we have $dB/dr = 0$, with the result that at point 1 the force is zero. This also follows from the fact that point 1 in the middle of the solenoid is the equilibrium point of the contour positioned inside the solenoid. At point 2 $(r = a)$,

$$\frac{dB}{dr} = \frac{\mu_0 I N_0}{2R} \left[\frac{1}{(1 + 4a^2/R^2)^{3/2}} - 1 \right], \qquad (5.12.3)$$

while at point 3 $(r = 2a)$,

$$\frac{dB}{dr} = \frac{\mu_0 I N_0}{2r} \left[\frac{1}{(1 + 9a^2/R^2)^{3/2}} - \frac{1}{(1 + a^2/R^2)^{3/2}} \right]. \qquad (5.12.4)$$

Comparison of (5.12.3) and (5.12.4) shows that the numerical value of the derivative is greater at point 2 than at point 3. It can also be verified that at all points outside the solenoid the attractive force (if the direction of the current in the contour is opposite to that of the current in the solenoid, the force is repulsive) is smaller than at an end face of the solenoid, and decreases as the distance from the solenoid grows.

Formula (5.12.1) can be obtained in the following manner (see Figure (b) accompanying the answer). The element dx of the length of the solenoid contains $N_0\,dx$ turns (with N_0 the number of turns per unit length). The induction at point A generated by the current flowing in these turns is

$$dB = \frac{\mu_0}{4\pi}\,\frac{I N_0\,dx\,R\,d\varphi}{z^2},$$

where $R\,d\varphi$ is the element of length of the turn subtended by an angle $d\varphi$. The projection of dB on the solenoid axis is

$$dB_\parallel = \frac{\mu_0 I N_0\,dx\,R\,d\varphi}{z^2}\cos\alpha. \qquad (5.12.5)$$

The perpendicular component of dB is compensated by the induction generated by the symmetric elements of the same turn. Expressing all quantities in terms of angle α and the solenoid radius R, we get

$$z = \frac{R}{\cos\alpha}, \quad r - x = R\tan\alpha, \quad dx = -\frac{R}{\cos^2\alpha}\,d\alpha.$$

Substituting all this into (5.12.5), we find that

$$dB_\parallel = -\mu_0 I N_0\,d\varphi\cos\alpha\,d\alpha,$$

239

which yields formula (5.12.1) after we integrate from 0 to 2π with respect to φ and from α_1 to α_2 with respect to α.

5.13. Initially the external magnetic flux coupled with the contour is zero. In tending to increase this flux, the contour moves in such a manner that (a) the magnetic moment vector associated with the contour aligns with the induction vector of the external field and (b) **contour** moves into the region of higher induction after the **alignment** is completed. Under the given directions of the currents, the induction generated by the solenoid is directed from right to left and the magnetic moment vector of the contour is directed upward. Thus, the contour rotates counterclockwise and moves toward the solenoid. If the direction of current in the contour is opposite to the one shown in the figure accompanying the problem, the contour rotates clockwise and also moves toward the solenoid.

5.14. Contour *2* is in a nonuniform magnetic field. If the current in this contour flows in the same direction as the currents in contours *1* and *3*, contour *2* is attracted to the other two contours. If it is deflected from the state of equilibrium in some direction, then from this direction there acts on it an attractive force that is greater than the other attractive force (since contour *2* is in a nonuniform magnetic field), and this means that it will move in that direction and will be drawn closer to the corresponding contour. If the current in contour *2* flows in the direction opposite to that of the currents in contours *1* and *3*, then it might seem that contour *2* is in a state of stable equilibrium, since repulsive forces act from both directions. But there is another reason for instability. For an arbitrarily small rotation of contour *2* there appears a torque acting on this contour, and this torque tends to rotate the contour into such a position in which the direction of the current in contour *2* coincide with that of the currents in contours *1* and *3*. When this process is completed, we again have to deal with the instability considered in the first case.

Analyzing the behavior of contour *2*, we see that in both cases the instability manifests itself through a general rule, according to which a contour moves in an external field or changes its form in such a manner that the magnetic flux coupled with the contour acquires maximal value.

240

5.15. The work performed in moving a contour carrying a current is equal to $A = I\,(\Psi_2 - \Psi_1)$. If the flux coupled with the contour whose position is changed is initially Ψ_1, then upon rotating the plane of the contour by 180° this flux becomes $-\Psi_1$, upon rotating the plane by 90° it drops to zero, and upon moving the contour whose position is changed away from the fixed contour the flux decreases but does not become zero. Thus, in the first case $A_1 = -2I\Psi_1$, in the second $A_2 = -I\Psi_1$, and in the third $A_3 = -I\,(\Psi_1 - \Psi_2)$, where Ψ_2 is the flux coupled with the contour upon moving the mobile contour away from the fixed contour. The minus that is present in each formula shows that the work must be done against the interaction of the contours.

5.16. The velocity of each particle may be decomposed into two components, one pointing along the induction vector, and the other at right angles to the induction vector. The component directed along the field does not change since the Lorentz force affects only the component that is perpendicular to the field. If we denote this latter component by $v_\perp$ the Lorentz force is

$$F = ev_\perp B. \qquad (5.16.1)$$

This force, which is perpendicular both to the velocity of a charged particle and to the induction vector, imparts a normal acceleration to the particle in question, with the equation of motion of the particle in the direction perpendicular to the field being

$$mv_\perp^2/R = ev_\perp B. \qquad (5.16.2)$$

Combining (5.16.1) with (5.16.2), we can determine the radius of the circle along which the particle moves and the time it takes the particle to complete one circle (which does not depend on the velocity). In the course of the same time interval, T, the particle also moves along the field by a distance $h = v_{||}T$, where $v_{||}$ is the component of the velocity along the field. The result is the motion of the particle along a helical line with radius R and lead h. Since for an initial velocity v and an angle α the longitudinal component of the velocity is $v_{||} = v \cos \alpha$ and the transverse component is $v_\perp = v \sin \alpha$, the trajectory of the particle with the larger angle α has a greater radius and a smaller lead of the helical line.

5.17. An electron accelerated by a potential difference U acquires kinetic energy

$$mv^2/2 = eU. \qquad (5.17.1)$$

The force acting on the electron in a magnetic field is the Lorentz force

$$F = evB,$$

which makes the electron move along a circular arc whose radius is R, so that, according to Newton's second law,

$$mv^2/R = evB. \qquad (5.17.2)$$

The induction of the magnetic field generated by the current in the solenoid is

$$B = \mu_0 I N_0. \qquad (5.17.3)$$

Excluding velocity v from Eqs. (5.17.1) and (5.17.2) and substituting the value of B from (5.17.3), we find the sought-for charge-to-mass ratio:

$$\frac{e}{m} = \frac{2U}{\mu_0^2 I^2 N_0^2 R^2}.$$

5.18. The electric field vector inside the capacitor is directed at right angles to the capacitor plates. The force $F_e = QE = QU/l$ with which the electric field acts on the particle is directed in the same manner. A force equal in magnitude to F_e but pointing in the opposite direction acts on the particle from the magnetic field. According to the Lorentz formula, this force is $F_m = QvB$ and is directed at right angles to the velocity of the particle and the magnetic induction vector. This means that the induction vector must be perpendicular to the electric field. As we have said earlier, the two forces must be equal: $QU/l = QvB$, or

$$B = U/(vl). \qquad (5.18.1)$$

The velocity the particle acquired in an electric field can be found by employing the energy conservation law:

$$mv^2/2 = QU_0.$$

Solving this for v and substituting the result into (5.18.1), we finally obtain

$$B = \frac{U}{l} \sqrt{\frac{m}{2QU_0}}.$$

5.19. According to the Lorentz formula, charges moving in a magnetic field are subjected to a force whose direction is determined via the left-hand rule, where the positive direction of a current is defined in the "electrical-engineering" sense, that is, the direction in which the positive charges move in the conductor. Therefore, irrespective of the sign of the charge carriers, the forces acting on these carriers point in the same direction. In the case illustrated in the figure accompanying the problem, the charges move downward. In a metal or an n-type semiconductor, where electrons are the charge carriers, this will result in a depletion of charge carriers in the region about point a; the region will acquire a positive potential. In the case of a p-type semiconductor the sign of the charge is obviously minus.

5.20. According to Lenz's law, the induced current is in such a direction as to oppose the change in the magnetic field that produces it (that is, oppose the change in magnetic flux coupled with the contour). When the two contours approach each other, the flux coupled with the second contour increases, which means that the direction of the induction current in that contour is opposite to the current in the contour. On the other hand, when the contours are moved away from each other, the decrease in the flux in contour *2* leads to an induction current in that contour that is directed in same sense as in contour *1*.

5.21. The induction emf is

$$\mathcal{E}_i = -\frac{d(LI)}{dt} = -L\frac{dI}{dt} - I\frac{dL}{dt}.$$

In the case at hand the variable quantity is the inductance. When the spiral is stretched, the inductance falls, so that $dL/dt < 0$ and $\mathcal{E}_i > 0$. The generated induction emf leads to an increase in the current in the circuit. For an exact calculation one is forced to solve the equation

$$I = \left(\mathcal{E}_0 - L\frac{dI}{dt} - I\frac{dL}{dt} \right) \Big/ R,$$

which requires knowing the time dependence of the inductance, $L = L(t)$.

5.22. Since removal of the iron core results in a decrease in the induction and the magnetic field flux in the solenoid, during removal there emerges a self-induction emf, which opposes the reduction in the flux and, hence, increases the current flowing in the solenoid (the direction of the external emf, which supplies DC power to the solenoid, and that of the self-induction emf are the same).

5.23. The self-induction emf defined by the formula $\mathscr{E}_{s.i} = -L(dI/dt)$ is proportional to the derivative dI/dt (for equal inductances), which is the greater the steeper the straight line. Hence, the self-induction emf is higher for the inductance for which the time dependence of the current is depicted by straight line *1*. Since the slopes of the straight lines do not change when the currents pass through zero, both the numerical values and the directions of the self-induction emf's are retained.

5.24. The self-induction emf defined by the formula

$$\mathscr{E}_{s.i.} = -L\frac{dI}{dt}$$

has its maximal value, obviously, at the point where the rate of decrease of current is greatest, that is, at point *3*.

5.25. In Figure (a), after key K is closed, the current flowing through the circuit that consists of L and R connected in series is initially the same as the current that was flowing before K was closed. For this circuit we can write Kirchhoff's law in the form

$$-L\frac{dI}{dt} = RI.$$

Separation of variables and subsequent integration yield

$$I = I_0 \exp(-Rt/L).$$

The current falls off according to an exponential law, with the self-induction emf being initially

$$\mathscr{E}_{s.i.} = -L\frac{dI}{dt} = I_0 R,$$

which means that the self-induction emf is equal to the emf of the DC source.

244

In Figure (b), after key K is opened, the current initially is the same as the one that was flowing in the circuit before K was opened. In this case, however, ther esistor R closes the circuit. Since prior to opening the key the current flowing in the resistor was much weaker than that flowing in the induction coil, the voltage across the resistor after K is opened may initially become considerably higher than that prior to opening the key, which is possible only if R is considerably higher than the resistance of the DC source. One must bear in mind also that after opening the key the current in the resistor will reverse its direction.

5.26. The increase in current in the circuit with a resistance and an inductance occurs according to the law

$$I = \frac{\mathcal{E}}{R} [1 - e^{(R/L)t}]. \qquad (5.26.1)$$

Since by hypothesis only one parameter can vary, the parameter may be only the inductance because conservation of the limiting current is possible only when two parameters, $\mathcal{E}$ and R, are varied simultaneously. Formula (5.26.1) implies that the increase in current is the slower the higher the inductance in the circuit. Hence, curve 2 corresponds to a higher inductance.

5.27. The magnetic flux coupled with contour 2 is

$$\Psi = BS,$$

where S is the area of the contour, and B is the magnetic induction at the point where the contour has been placed. Accordingly, the induction emf generated in the contour is

$$\mathcal{E}_1 = -S = \frac{dB}{dt} - S \frac{dB}{dr} \frac{dr}{dt} = -Sv \frac{dB}{dr}.$$

The induction of the magnetic field generated by the current flowing in a circular contour and measured at a certain distance from the contour on the contour's axis is given by the formula

$$B = \frac{\mu_0 p_m}{2\pi (R^2 + r^2)^{3/2}},$$

where p_m is the magnetic moment of the contour. Therefore, the induction emf in contour 2 is

$$\mathcal{E}_1 = \frac{3\mu_0 p_m Sv}{2\pi} \frac{r}{(R^2 + r^2)^{5/2}} = C \frac{r}{(R^2 + r^2)^{5/2}},$$

245

with $C = 3\mu_0 p_m Sv/2\pi$. The sign of r determines the sign of the induced emf; when contour 2 is moved closer to contour 1, r is negative and so is the induced emf, so that the direction of the current in contour 2 is opposite to that of the current in contour 1. When contour 2 passes through contour 1, the induced emf changes sign. The maximal value of this emf can be obtained by nullifying the derivative,

$$\frac{d\mathscr{E}_1}{dt} = C\,\frac{(R^2+r^2)^{5/2}-(5/2)\,(R^2+r^2)^{3/2}\,2r^2}{(R^2+r^2)^5} = 0.$$

Thus the emf is maximal at $r = R/2$. An emf of equal magnitude but of opposite sign is generated at the same distance but on the other side when the contours are brought together.

5.28. When key $K1$ is closed, a closed circuit consisting of a DC source and the induction coil $L1$ is formed.

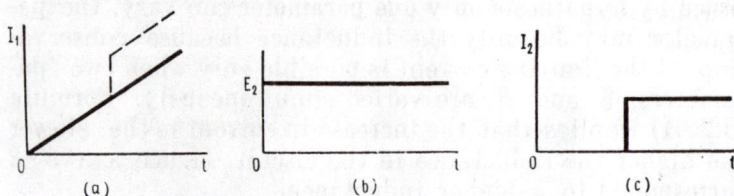

(a) (b) (c)

Fig. 5.28

Since there is no resistance in this circuit, the sum of the emf's is zero:

$$\mathscr{E} - L_1 \frac{dI_1}{dt} = 0.$$

The fact that $\mathscr{E}$ and L_1 are constant requires that dI_1/dt be constant, too. Thus, a current linearly increasing with time will flow in the circuit (the solid line in Figure (a) accompanying the answer). The magnetic flux generated by this current is coupled with both coils and also linearly increases with time:

$$\Psi = L_1 I_1.$$

In the second coil there appears a constant emf (Figure (b) accompanying the answer) whose direction is opposite to that of the current in $L1$:

$$\mathscr{E}_2 = -\frac{d\Psi}{dt} = -L_1 \frac{dI_1}{dt} = -L_2 \frac{dI_1}{dt}.$$

The current in the first coil increases with a constant time rate as long as *K2* is open. When *K2* is closed, a current flows in *L2*, with the magnetic field generated by this current opposing the field generated by the current in *L1*. The fact that the instantaneous value of the flux coupled with both coils must be preserved requires that there be a jump in the current in *L1*, after which the current will continue to grow according to the same linear law (the dashed broken line in Figure (a)). The current in *L2* will remain constant during the entire process (Figure (c) accompanying the answer).

5.29. (1) To find the mutual inductance we determine the magnetic flux coupled with the contour and generated by current I flowing in the straight conductor. In this case the mutual inductance is determined from the equation $M = \Psi/I$. The induction at a distance x from the straight conductor is $B = \mu_0 \mu I / 2\pi x$. The fluxes that flow through a part of the contour dx wide and b high and through the entire contour are, respectively,

$$d\Phi = \frac{\mu_0 \mu}{2\pi} \frac{I}{x} b \, dx,$$

$$\Phi = \frac{\mu_0 \mu}{2\pi} I b \int_c^{c+a} \frac{dx}{x} = \frac{\mu_0 \mu}{2\pi} I b \ln \frac{c+a}{c}.$$

When there are N turns in the contour, the flux coupled with the contour is

$$\Psi = N\Phi = \frac{\mu_0 \mu}{2\pi} I N b \ln \frac{c+a}{c},$$

which implies that the mutual inductance is

$$M = \frac{\mu_0 \mu}{2\pi} b N \ln \frac{c+a}{c}.$$

(2) Since the rotation of the contour through 90° makes the flux coupled with the contour vanish, the amount of electricity induced in the contour as a result of such a rotation is determined by the formula

$$Q = \frac{\Psi}{R} = \frac{\mu_0 \mu b N I}{2\pi R} \ln \frac{c+a}{c}.$$

(3) The rotation of the contour through 180° requires the following work to be done:

$$A = -2\Psi I = -\frac{\mu_0 \mu I^2 N b}{\pi} \ln \frac{c+a}{c}$$

247

(since after the rotation the flux coupled with the contour will become $-\Psi$). The "minus" shows that the work is done against the forces induced by the magnetic field.

5.30. When a current flows in a conductor, the induction of the magnetic field generated by this current at a distance r from the conductor is

$$B = \mu_0 I/2\pi r.$$

The magnetic flux coupled with the contour formed by the winding of the loop is $\Psi = BSN$, or

$$\Psi = \mu_0 ISN/2\pi r.$$

When the current drops to zero, the flux follows it, and the amount of electricity flowing in the contour is determined by the formula $Q = \Delta\Psi/R$. Hence, the current that had been flowing in the conductor prior to switch-off is

$$I = \frac{2\pi rRQ}{\mu_0 SN}.$$

5.31. The following emf is induced in the coil:

$$\mathscr{E}_1 = -\frac{d\Psi}{dt}. \tag{5.31.1}$$

We see that the maximal possible value of $\mathscr{E}_i$ is the higher the greater the rate with which the coil is moved out of the field. The area under the curve is given by the integral

$$\int_0^t \mathscr{E}_1\, dt = -\int_{\Psi_1}^{\Psi_2} d\Psi = \Psi_1 - \Psi_2 = \Psi_1 = BSN$$

and, hence, is independent of the rate of coil removal from the region with the field.

5.32. (1) The system can be considered as being a new solenoid whose length is twice as large as that of one solenoid, with a density of turns the same as that in one solenoid and with the same cross-sectional area. Since the inductance of one solenoid is $L = \mu_0 N_0 V$, where N_0 is the number of turns per unit length, and V is the solenoid volume, which in this case is twice the volume of one solenoid, we have $L_1 = 2L_0$. The same result can be achieved by considering the self-induction emf that is generated in the two solenoids connected in series. The total emf is equal to the sum of the emf's generated in each solenoid; hence,

$$\mathcal{E}_1 = -2L_0 \frac{dI}{dt},$$

which yields $L_1 = 2L_0$.

(2) When the solenoids are connected in parallel, the self-induction emf in each solenoid is

$$\mathcal{E}_1 = -L_0 \frac{d(I/2)}{dt} = -\frac{1}{2} L_0 \frac{dI}{dt}.$$

Because the solenoids are connected in parallel, the total emf has the same value. Thus, with a current I in a circuit that is external with respect to the solenoid, the induced emf is one-half the value for the inductance L_0. Hence,

$$L_2 = \frac{1}{2} L_0.$$

(3) In this case, the number of turns per unit length is twice as large as that of one solenoid, and since the inductance is proportional to N_0^2, we have (provided that the current remains unchanged)

$$L_3 = 4L_0.$$

(4) If one solenoid is fitted onto the other and the senses of the turns coincide and the solenoids are connected in parallel, the current through each solenoid is $I/2$ if the current in the circuit is I, while the flux associated with current $I/2$ is $\Phi/2$. The total flux is Φ and the flux coupled with each solenoid is $\Psi = \Phi N_0$. In each solenoid there is generated a self-induction emf equal to the one induced in a separate solenoid when current I varies. Since the solenoids are connected in parallel, this emf is the common emf of both solenoids. Hence,

$$L_4 = L_0.$$

(5, 6) In both cases the induction flux in the solenoids is zero, so that $L_5 = L_6 = 0$.

5.33. The induced emf is

$$\mathcal{E}_1 = -L \frac{dI}{dt}.$$

Hence, the value of the emf is determined by the rate with which the current decreases (the sign of this rate is opposite to dI/dt). The slope of the straight line on the *0-1* section is twice as large as that of the straight line on the *1-2* section and coincides (numerically) with the slopes of the straight lines on sections *3-4-5* and *5-6*. Hence, in the time interval between points *1* and *2* the

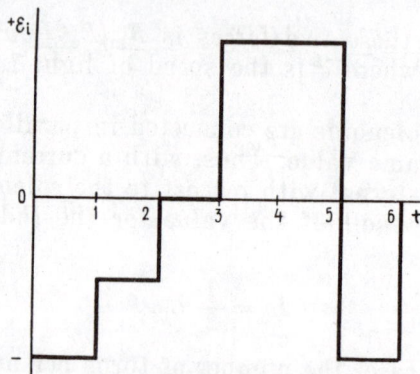

Fig. 5.33

induced emf is one-half of the emf's in the other intervals except the interval from point *2* to point *3* where $\mathscr{E}_1 = 0$ ($I = \text{const}$).

5.34. In each beam we isolate an element of length l. On the one hand, the element can be thought of as a charge $Q = enSl$, or, on the other, as an element of current $I = envS$. An electrostatic repulsive force $F_e = EQ$ acts on each charge element, where E can be assumed to be the electric field generated by an infinitely long straight conductor carrying a charge whose linear density is $\tau = enS$. This field, which acts on the charges in the second beam, can be written in the form

$$E = \frac{enS}{2\pi\varepsilon_0 r} ,$$

so that

$$F_e = \frac{e^2 n^2 S^2 l}{2\pi\varepsilon_0 r} .$$

The isolated element, if considered as an element of current, is under a force $F_m = BIl$, where B is the induction generated by the other current:

$$B = \mu_0 \frac{envS}{2\pi r}.$$

Thus

$$F_m = \mu_0 \frac{e^2 n^2 v^2 S^2 l}{2\pi r}.$$

The ratio of these two forces is $F_m/F_e = v^2 \varepsilon_0 \mu_0$. Since $\varepsilon_0 \mu_0 = 1/c^2$, where c is the speed of light in vacuo, we obtain

$$\frac{F_m}{F_e} = \frac{v^2}{c^2}.$$

5.35. The reasoning is all wrong. Even if the electrons in the conductors are at rest in relation to the needle, the positive ions that are moving in this case in the opposite direction create, obviously, a magnetic field equal to the one generated by the moving electrons when the needle was at rest. If the electrons are moving in a vacuum, then the electrodes and the electric field move in the opposite direction (when the needle is at rest in relation to the electrons).

5.36. The permeability of air is practically unity and at any point the magnetic field vector coincides in direction with the magnetic induction vector. In the emu system of units both vectors coincide in magnitude as well, while in the SI system they are related thus: $H = B/\mu_0$. Since the lines of force of induction are continuous, inside a bar magnet they are directed from the south pole to the north pole and are continued outside the magnet by lines directed from the north pole to the south pole. To determine the direction of the magnetic field inside the magnet, one must bear in mind that the circulation integral of the magnetic field vector along a closed contour must be equal to the algebraic sum of the currents encompassed by the contour. Since in the case at hand there are no currents, the circulation integral along any con-

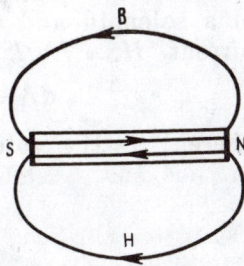

Fig. 5.36

251

tour lying inside the magnet must be zero. If the contour passes partially through the air surrounding the magnet and partially in the magnet, the circulation integral may be equal to zero only if inside the magnet the magnetic field vector is directed from the north pole to the south pole. Formally this means that inside the magnet the permeability is negative.

5.37. Alternating magnetization results in liberation of heat in the steels, with the amount of heat proportional to the area bounded by the hysteresis loop. Since a transformer operates on alternating currents, the amount of heat liberated in the core of a transformer will be the greater the bigger the area bounded by the loop. From this fact one can conclude that the steel whose hysteresis loop is depicted in Figure (b) accompanying the problem is more desirable. On the other hand, it is desirable that a permanent magnet have as high a residual magnetic induction and a coercive force as possible. This implies that the steel more suitable for manufacturing a permanent magnet is the one whose hysteresis loop is depicted in Figure (a).

5.38. The elementary work involved in changing the magnetic flux coupled with a contour carrying a current I is

$$dA = I \, d\Psi, \quad \text{or} \quad dA = ISN \, dB.$$

If we use the relationship that exists between the current in a solenoid and the magnetic field generated by this current, $H = IN/l$, we obtain

$$dA = H \, Sl \, dB = VH dB,$$

where V is the volume of the core. The entire work is

$$A = V \int_0^B H dB.$$

The integral on the right-hand side is the area bounded by the B vs. H curve, the ordinate, and the segment of a straight line parallel to the H axis (see the figure accompanying the answer).

5.39. As shown in the answer to Problem 5.36, the magnetic field inside a permanent magnet is directed from the north pole to the south pole, while the induction is directed from the south pole to the north pole, with the result that these quantities have opposite signs. On the hysteresis loop, this condition is met on sections *2-3* and *5-6* (see the figure accompanying the problem). Formally, to the permeability on these sections we can assign a negative value. Correspondingly, to point *0* we may assign a permeability equal to $\pm\infty$, while points *3* and *6* correspond to zero values of the permeability.

5.40. After the current in the conductor has ceased, the circulation integral of the magnetic field strength along any closed contour is zero (this is true even for a closed

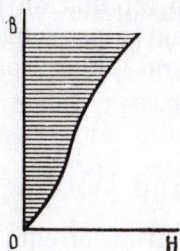

Fig. 5.38

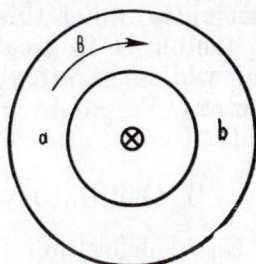

Fig. 5.40

contour that passes in the toroid). Since all points of a contour that is a circle concentric with the section of the conductor are identical, the magnetic field strength at all points inside the toroid is zero, too. At the same time, the toroid carries a residual magnetic induction whose lines of force are circles directed in the manner shown by the arrow in the figure accompanying the answer. This magnetic state of the toroid corresponds to point *2* or *5* on the hysteresis loop (the choice of the point depends on which of the two directions is assumed to be positive). If the positive direction of the magnetic field vector is the one the toroid acquires during magnetization, this magnetic state of the toroid is depicted by point *2*.

5.41. The circulation integral of the magnetic field is uniquely determined by the current flowing inside the contour. Because of this, the circulation integrals along contours *1*, *4* and *5* (see the figure accompanying the problem) are the same and equal to the current *I*, while the circulation integrals along contours *2* and *3* are zero. However, the situation with the circulation integrals of the magnetic induction along these contours is quite different. When the circulation integrals are evaluated along contours that pass through a homogeneous medium (in the case at hand, in a vacuum), they do not depend on the shape and size of the contours, with the result that the circulation integrals along contours *4* and *5* are equal. Reasoning in the same manner, we conclude that the circulation integral along contour *3* is zero. But in evaluating the circulation integrals along contours *1* and *2* that include sections of a medium with a permeability greater than unity, the circulation elements in this medium are μ times greater than in the vacuum (if μ is greater than unity). For this reason the circulation integral along contour *1* is greater than that along contours *4* and *5*, while the circulation integral along contour *2* is nonzero.

6. Oscillatory Motion and Waves

6.1. Equal deflections from the position of equilibrium occur if

$$\sin \Phi_1 = \sin \Phi_2, \qquad (6.1.1)$$

where $\Phi_1 = \omega t_1$ and $\Phi_2 = \omega t_2$ (as shown in the figure, the initial phase is zero). The x vs. ωt curve shows that condition (6.1.1) is met if

$$\sin \Phi_2 = \sin (\pi - \Phi_1).$$

Here $\cos \Phi_2 = -\cos \Phi_1$, that is, phases Φ_1 and Φ_2 correspond to velocities of the oscillating point that are

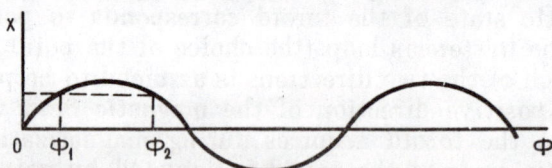

Fig. 6.1

opposite in direction. The phases of harmonic oscillations coincide if both the deflections and the velocities of the oscillations coincide (both in absolute value and in direction).

6.2. The amplitude of the oscillations depicted by curve *2* in the figure accompanying the question is twice as large as that of the oscillations depicted by curve *1*.

The periods of the two oscillations coincide. Oscillations 2 lag in phase behind oscillations 1 by a quarter of one period. Hence, oscillations 2 are represented by the equation

$$x = 2A \sin (\omega t - \pi/2).$$

6.3. Oscillations 1 have a period that is twice as large as that of oscillations 2, so that the frequency of oscillations 1 is one-half of that of oscillations 2. Amplitude A_1 is twice as large as amplitude A_2. The energies of these oscillations are

$$W_1 = \frac{1}{2} m\omega_1^2 A_1^2$$

and

$$W_2 = \frac{1}{2} m\omega_2^2 A_2^2 = \frac{1}{2} m \left(\frac{\omega_1}{2} \right)^2 (2A_1)^2 = W_1,$$

that is, coincide.

6.4. The equation of the motion projected on the x axis is

$$x = A_x \sin \omega t.$$

In the case where the object moves clockwise, the deflection along the y axis at time zero $(t = 0)$ is $y = A_y$, and then it decreases to zero when the maximum on the x axis is attained. The sine decreases from unity to zero as the angle changes from $\pi/2$ to π. In this case the initial phase of oscillations along the y axis is $\pi/2$, and the equation of motion projected on the y axis is

$$y = A_y \sin (\omega t + \pi/2).$$

In the case where the object moves counterclockwise, the deflection along the y axis is zero when the phase of motion along the x axis becomes $\pi/2$ and, hence, the initial value of this deflection is $y = - A_y$ and increases to zero in the course of a quarter of the period. In the case at hand the equation of motion projected on the y axis can be written in the form

$$y = A_y \sin (\omega t - \pi/2).$$

6.5. In the first case the oscillations along the y axis begin $\pi/2$ earlier in phase than along the x axis, while in the second case they lag behind by the same quantity.

In both cases the motion takes place along an ellipse described by the equation

$$\frac{x^2}{A_x^2} + \frac{y^2}{A_y^2} = 1.$$

The two motions differ in direction. In the first case the motion is clockwise while in the second it is counterclockwise. The equations of motion have the form

$$x = A_x \sin \omega t, \quad y_1 = A_y \sin\left(\omega t + \frac{\pi}{2}\right), \quad y_2 = A_y \sin\left(\omega t - \frac{\pi}{2}\right).$$

6.6. When the deflection along the x axis is zero and the velocity is positive, the deflection along the y axis is greater than zero but smaller than A_2, with y continuing to increase according to the direction designated by the arrow and reaching the value A_2 (i.e. when $\omega t + \varphi = \pi/2$) for $0 < \omega t < \pi/2$. Hence,

$$0 < \varphi < \pi/2.$$

6.7. In the course of one period the oscillating point attains each of its maximal (but opposite) values once (i.e. in the motion along an axis). For this reason the complete Lissajous figure touches the sides of the rectangle limiting the motion exactly the same number of times as there are periods in the motion of the point in a certain direction. Along the x axis the figure touches the sides of the rectangle twice, while along the y axis four times. Hence

$$\omega_2 = 2\omega_1 \quad \text{and} \quad y = A_2 \sin(2\omega_1 t + \varphi).$$

To determine φ, we assign to $\omega_1 t$ the values that correspond to points where the Lissajous figure touches the limiting rectangle. For instance, if we take $\omega_1 t = \pi/2$, then $2\omega_1 t + \varphi = \pi/2 + \varphi$. Here

$$\sin(2\omega_1 t + \varphi) = -1.$$

Hence,

$$\pi/2 + \varphi = -\pi/2, \quad \text{or} \quad \varphi = -\pi.$$

6.8. Just like in the previous problem, the number of periods it takes to traverse completely the Lissajous figure in either direction is determined by the number of points where the Lissajous figure touches the rectangle that limits the motion. There are three such points in the positive direction of x and two points in the positive direction of y. Thus, the entire figure is traversed in the direction x in the course of three periods and in the direction y in the course of two periods. Hence,

$$\omega_1/\omega_2 = 3/2.$$

6.9. The kinetic energy is maximal when the velocity is maximal in absolute value. Being the time derivative of displacement, the velocity is maximal at moment *2*. The maximal potential energy is determined by the maximal displacement, that is, the amplitude, and is equal to $kA^2/2$. Hence, it is maximal at moment *1*. At this moment the kinetic energy is zero, while the potential energy is zero at moment *2*. The acceleration of the particle is at its maximum when the second time derivative of the displacement is maximal. This corresponds to moment *1*. Since at this moment the second derivative is negative, so is the acceleration.

6.10. The period of harmonic oscillations that take place due to a quasielastic force $(F = -kx)$ is determined from the formula

$$\tau = 2\pi \sqrt{m/k}. \qquad (6.10.1)$$

The spring constant k is defined as the force that is required to stretch the spring in such a manner that the spring elongation becomes equal to its initial length. In the case at hand the elongations occur because of the weight of the loads, with the result that

$$k_1 = m_1 g/l \quad \text{and} \quad k_2 = m_2 g/l.$$

Susbtituting k into (6.10.1), we see that the masses cancel out and in both cases the period is

$$\tau = 2\pi \sqrt{l/g}.$$

The same result can be obtained (to within a constant coefficient) from dimensional reasoning. There are three quantities that appear in the problem: mass, elongation, and time (the sought period). In addition, since forces equal to the weights of the loads are applied to the springs, we may assume that the acceleration of gravity g will enter into the solution. Bearing in mind that the dimensions of the left- and right-hand sides of any equation must be the same, we can write

$$T = M^a L^b [LT^{-2}]^c,$$

where a, b, and c are the exponents of the corresponding quantities. We have the following equations for the exponents:

$$a = 0, \quad b + c = 0, \quad c = -1/2.$$

Hence,

$$\tau = \mathcal{K}\, l^{1/2} g^{1/2},$$

where $\mathcal{K}$ is a dimensionless coefficient, which cannot be found using solely dimensional considerations. Above it was shown that this coefficient is equal to 2π.

The energy of the oscillations of a load can be written in the form

$$W = mA^2\omega^2/2.$$

Since the periods of oscillations (and hence the frequencies) are equal and so are the amplitudes (by hypothesis), the load with the higher energy is the one whose mass is m_1.

6.11. In the case at hand the quasielastic force is Archimedes' force. When the bottom of the test tube lies above or below the position of equilibrium by a distance x, this force is

$$F = -Sx\rho g. \tag{6.11.1}$$

The mass of the test tube together with the mass of the load is equal to the mass of the displaced water, or

$$m = lS\rho. \tag{6.11.2}$$

Using (6.11.1), we can find the "spring constant"

$$k = |F|/x = S\rho g. \tag{6.11.3}$$

Substituting (6.11.2) and (6.11.3) into the expression for the period of oscillations (6.10.1), we get

$$\tau = 2\pi \sqrt{m/k} = 2\pi \sqrt{l/g}.$$

We see that τ depends neither on the mass and cross-sectional area of the tube nor on the density of the liquid. The same result can be obtained from dimensional considerations, just like it was done in Problem 6.10.

6.12. If m_0 is the known mass and m is the unknown mass and if ω_0 and ω are the angular frequencies of oscil-

258

lations of the systems with the known mass and the known mass plus the unknown, then

$$\omega_0 = \sqrt{k/m_0}, \qquad (6.12.1)$$

$$\omega = \sqrt{k/(m_0 + m)}, \qquad (6.12.2)$$

where k is the spring constant. Combining (6.12.1) with (6.12.2), we arrive at a formula for the unknown mass:

$$m = m_0 \left(\frac{\omega_0^2}{\omega^2} - 1 \right).$$

6.13. The total energy of oscillations of a material particle can be made equal to the maximal kinetic energy or maximal potential energy of the particle. In the case at hand it proves expedient to compare the maximal potential energies, which are specified by the maximal deflections. When the deflection is at its maximum, the load (or particle) is at a height h above the position of equilibrium:

$$h = l \, (1 - \cos \alpha).$$

Since the expression inside the parentheses is the same for both pendulums, the pendulum with the greater length is raised to the greater height and, hence, has the higher energy.

6.14. Just like in the previous problem, the total energy can be made equal to the maximal potential energy. Since the center of gravity of the physical pendulum is higher than that of the simple pendulum, the physical pendulum can be thought of as a simple pendulum of smaller length. Thus, the given simple pendulum has a higher energy.

6.15. In the case at hand the disk constitutes a physical pendulum. The period of oscillations of a physical pendulum is given by the formula

$$T = 2\pi \left(\frac{J}{mgR_c} \right)^{1/2}.$$

The moment of inertia of the disk about the center is $J = mR^2/2$. According to Steiner's theorem,

$$J = m \, (R^2/2 + R_c^2),$$

whence

$$T = 2\pi \left[\frac{(R^2/2 + R_c^2)}{gR_c} \right]^{1/2}.$$

As expected, the period does not depend on the mass of the pendulum.

6.16. The angular frequency of oscillations for a physical pendulum is

$$\omega = (mgR_c/J)^{1/2},$$

where m is the mass of the pendulum, and J is the pendulum's moment of inertia. If the distance from the center of gravity to the point of suspension is R_c, then, according to Steiner's theorem, the moment of inertia of the rod about the suspension point is equal to the moment of inertia of the rod about the center of gravity plus the moment of inertia of a material particle whose mass is that of the rod about the point of suspension:

$$J = \frac{ml^2}{12} + mR_c^2.$$

Thus,

$$\omega = \left(\frac{12gR_c}{l^2 + 12R_c^2} \right)^{1/2}.$$

To find the extremum, we nullify the derivative of ω with respect to R_c:

$$\frac{d\omega}{dR_c} = \frac{6g\,(l^2 - 12R_c^2)}{R^{1/2}\,(l^2 + 12R_c^2)^{3/2}} = 0.$$

Whence

$$R_c = \frac{l}{2\sqrt{3}} = 0.29l.$$

6.17. The acceleration varies according to the same law as the force. Thus,

$$v = \frac{F_0}{m} \int\limits_0^t \sin \omega t \, dt = \frac{F_0}{m\omega} (1 - \cos \omega t) = v_m (1 - \cos \omega t).$$

The v vs. t curve is depicted in Figure (a) accompanying the answer. If the initial position of the point is taken as the origin, then

$$x = v_m \int\limits_0^t (1 - \cos \omega t) \, dt = v_m t - \frac{v_m}{\omega} \sin \omega t.$$

Thus, we have found that the particle is in translational motion with a velocity that periodically increases from zero to its maximum, equal to $2v_m$, and then drops off to

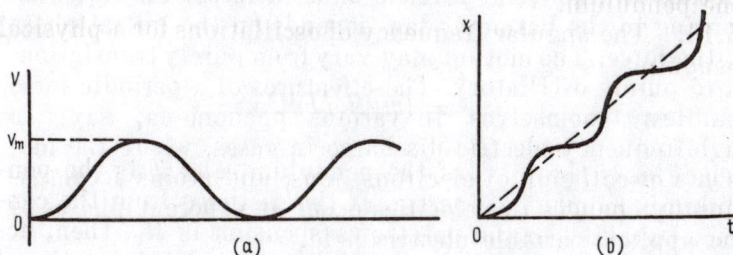

Fig. 6.17

zero. The motion is depicted schematically in Figure (b) accompanying the answer.

6.18. The solution to this problem is similar to that of Problem 6.17, the difference being that the initial phase

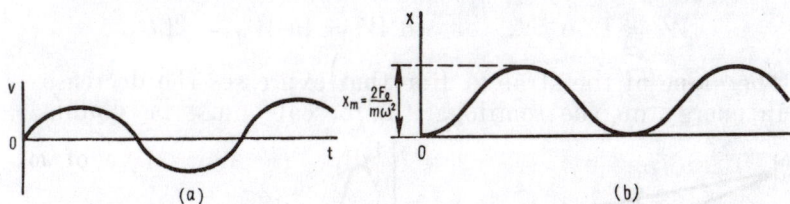

Fig. 6.18

of the driving force is different. In the case at hand, initially the force is maximal. The time dependence of the velocity is

$$v = \frac{F_0}{m} \int_0^t \cos \omega t = \frac{F_0}{m\omega} \sin \omega t = v_m \sin \omega t.$$

In contrast to the previous case, the velocity changes its direction during motion (Figure (a) accompanying the answer). The displacement of the particle can be found after integration:

$$x = v_m \int_0^t \sin \omega t = \frac{v_m}{\omega} (1 - \cos \omega t).$$

Thus, in the case at hand the motion is purely harmonic, as shown by the curve in Figure (b).

261

A comparison of the results of the previous problem with those of the present problem demonstrates that the motion of a material particle under a force that varies according to the harmonic law depends on the initial phase of the force. The motion may vary from purely translational to purely oscillatory. These features of a periodic force manifest themselves in various phenomena, say, in high-frequency electric discharge in gases, where the moments of collision of electrons, ions, and atoms accompanied by changes in velocities occur at different phases of the applied variable electric field.

6.19. If the amplitude decreases with the passage of time according to the law

$$A = A_0 e^{-\beta t},$$

then, since the oscillation energy is proportional to the square of the amplitude, the decrease in energy occurs according to the law

$$W = W_0 e^{-2\beta t}, \quad \text{or} \quad \ln W = \ln W_0 - 2\beta t.$$

The slope of the straight line that expresses the decrease in energy on the semilogarithmic scale must be double

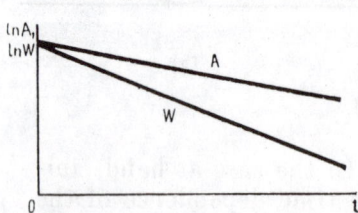

Fig. 6.19

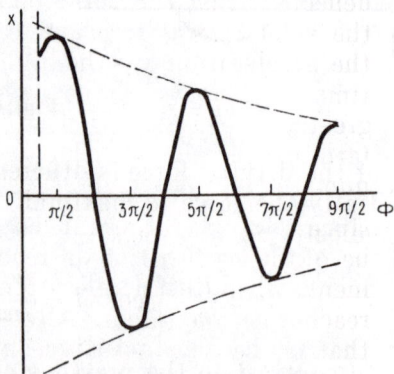

Fig. 6.20

the slope of the straight line that expresses the decrease in amplitude.

6.20. The figure accompanying the problem shows that the initial phase is $\pi/4$ while the ratio of the amplitude whose phases differ by 2π is equal to 1.5. This means that the logarithmic decrement $\ln (A_{n+1}/A_n)$ is approximately equal to 0.4.

6.21. Initially the velocity of the pendulum is zero and tends to zero as the pendulum approaches its position of equilibrium, so that it first grows and then, after passing through its maximum, decreases. We can arrive at the same conclusion after analyzing qualitatively the differential equation of the motion of the pendulum written

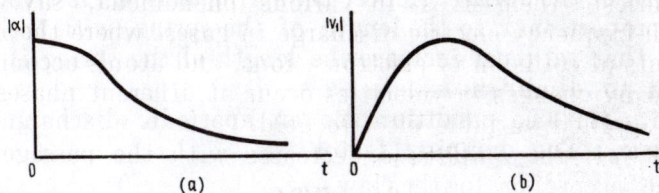

Fig. 6.21

in polar coordinates in the common approximation of small deflections:

$$I\ddot{\alpha} = -q\alpha - r\dot{\alpha}.$$

We select a system of coordinates in which the positive direction is the one in which the pendulum was initially deflected from the point of equilibrium. Initially, when the velocity was zero and the deflection was the largest, the acceleration was the highest. The curve depicting the time dependence of the deflection has at this point the greatest curvature. In the process of motion, the first term on the right-hand side of the equation decreases in numerical value, while the second term (which is positive since $\dot{\alpha} < 0$) grows, and because of this the absolute value of the acceleration decreases. There finally comes a moment when the acceleration vanishes and the velocity reaches its maximum. After that the acceleration grows, that is, becomes positive and increases in numerical value, which in the system of coordinates employed here implies deceleration, and the pendulum asymptotically approaches the position of equilibrium. The time dependences of the absolute values of the deflection and the velocity of the pendulum are shown in Figures (a) and (b) accompanying the answer.

6.22. In damped oscillations the damping factor is smaller than the natural frequency of free oscillations of the system: $\beta < \omega_0$. In aperiodic motion the situation is reversed: $\beta > \omega_0$. The damping factor is defined as the ratio $\beta = r/2m$, where r is the resistance of the medium, and m is the mass of the load. Both quantities remain unchanged, and so does β. To go over to the aperiodic mode, we must make ω_0 smaller. Since $\omega_0 = \sqrt{k/m}$, we must diminish k since m is fixed. At a given elongation force, the elongation of the spring is proportional to the initial length of the spring. Hence, the spring constant is inversely proportional to the length of the spring, with the result that we must increase the length of the spring if we wish to diminish k.

6.23. (1) The condition for an aperiodic discharge is $\beta > \omega_0$. The damping factor

$$\beta = R/2L \qquad (6.23.1)$$

does not depend on the capacitance. To make the process aperiodic, we must diminish the natural frequency, which for a fixed inductance means increasing the capacitance, and the easiest way to do this is to bring the plates of the capacitor closer together.

(2) According to (6.23.1), to decrease the damping factor for a fixed resistance, we must increase the inductance. To preserve the value of the natural frequency $\omega_0 = 1/\sqrt{LC}$, the capacitance must be decreased by the same factor. The frequency of the damped oscillations,

$$\omega = \sqrt{\omega_0 - \beta^2},$$

increases in the process, approaching ω_0.

(3) When the resistance and inductance are decreased simultaneously, the damping factor remains unchanged, but for a fixed capacitance the oscillation period $T = 2\pi/\sqrt{\omega_0^2 - \beta^2}$ decreases and, hence, so does the logarithmic decrement.

6.24. Both the logarithmic decrement and the period depend on the damping factor:

$$\theta = \beta T \qquad (6.24.1)$$
$$T = 2\pi/\sqrt{\omega_0^2 - \beta^2}. \qquad (6.24.2)$$

Since the lengths of the pendulums are equal, the natural frequencies of free oscillations (that is, without resistance) are equal, too. The damping factor is

$$\beta = r/2m, \qquad (6.24.3)$$

264

where r is the resistance of the medium, which is the same for the two pendulums. Substituting (6.24.3) into (6.24.1) and (6.24.2), we see that both the period and the logarithmic decrement of the sphere with the smaller mass are greater.

6.25. There is no periodic driving force in the system; hence, the oscillations are not forced. The oscillation frequency is determined by the mass and by the elastic properties of the spring, and since the amplitude of the oscillations remains unchanged, the oscillations are undamped although, of course, loss of energy is inevitable. This loss is compensated by the energy stored in the DC source. Thus, the oscillations belong to the type that occur with a natural frequency but with replenishing the energy from an external nonperiodic source, that is, self-oscillations.

6.26. The frequency dependence of the displacement amplitude in forced oscillations is given by the formula

$$A = \frac{F_0}{m \sqrt{(\omega_0^2 - \omega^2)^2 + 4\beta^2 \omega^2}} \, ,$$

while the frequency dependence of the velocity amplitude is given by the formula

$$v_\mathrm{m} = \frac{F_0 \omega}{m \sqrt{(\omega_0^2 - \omega^2)^2 + 4\beta^2 \omega^2}} \, .$$

In the first case, at $\omega = 0$ the amplitude A does not vanish but becomes equal to $F_0/m\omega_0^2$, or F/k, so that the curve cuts off a segment on the vertical axis, which segment is the displacement under a constant force. The velocity, of course, is zero in this case. Thus, the curves in Figure (a) correspond to the frequency dependence of the displacement amplitudes, while the curves in Figure (b) correspond to the frequency dependence of the velocity amplitudes. The smaller the damping factor β, the higher the curve in the respective diagrams. The damping factor also determines the position of the maxima of the displacement amplitudes:

$$\omega_\mathrm{res} = \sqrt{\omega_0^2 - 2\beta^2}.$$

The maximal velocity amplitude for all damping factors is achieved at $\omega = \omega_0$.

6.27. The displacement A_0 at $\omega = 0$ is determined by the ratio of the maximal force F to the elastic constant

265

k (the spring constant), or $A = F/k$. By hypothesis, both F_0 and k remain unchanged, whereby A does not depend on the resistance of the medium. The resonance frequency, defined as

$$\omega_{res} = \sqrt{\omega_0^2 - 2\beta^2},$$

is the closer to the natural frequency the smaller the values of the damping factor β. Since the latter is defined as $\beta = r/2m$ and the mass of the oscillating object remains unchanged (by hypothesis), β decreases and ω_{res} grows as r drops. The amplitude at the resonance frequency,

$$A_{res} = \frac{F_0}{2\beta m \sqrt{\omega_0^2 - \beta^2}},$$

is the higher the smaller the resistance of the medium.
6.28. The differential equation describing the behavior of the system is

$$m\ddot{x} + r\dot{x} + kx = F_0 \sin \omega t, \qquad (6.28.1)$$

and it has two solutions, a steady-state and a transient. The latter describes the process of setting in of forced oscillations. Usually only the steady-state solution is considered. However, at $r = 0$ and $\omega = \omega_0$ this equation has no steady-state solution, and because of this the amplitude continuously increases and so does the energy of the system, which energy is taken from the source of oscillations. In reality, a system in which the resistance of the medium is negligible for all practical purposes either behaves in such a manner that the amplitude reaches values at which Hooke's law ceases to be valid (and, respectively, Eq. (6.28.1) loses meaning) or is destructed. One must bear in mind also that the fact that we ignore the resistance of the medium, which at low velocities is a valid assumption, cannot be justified as the velocity grows higher and higher.
6.29. The resonance frequency is the same for both oscillations:

$$\omega_{res} = \sqrt{\omega_0^2 - 2\beta^2}.$$

Since the natural frequencies also coincide, so do the damping factors β. The resonance amplitude is

$$A_{res} = \frac{F_0}{2\beta m \sqrt{\omega_0^2 - \beta^2}}.$$

266

Only two quantities in this formula can vary: the mass of the oscillating object and the amplitude of the driving force. However, from the fact that the natural frequencies are the same and the damping factors are the same, it follows that for different masses only the elasticity coefficients and the resistances differ:

$$\omega_0 = \sqrt{k/m}, \quad \beta = r/2m.$$

But by hypothesis, the systems are supposed to differ only in one parameter. This parameter, therefore, can only be the amplitude of the driving force, which for one system is twice as high as for the other.

6.30. According to Huygens' principle, each point of a wavefront is an independent source of oscillations. An ap-

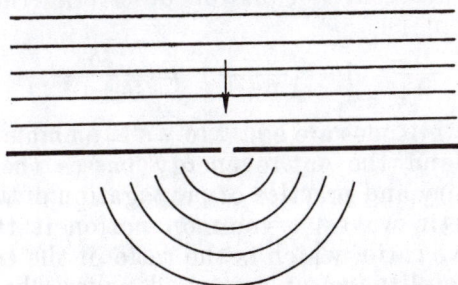

Fig. 6.30

erture whose width is much smaller than the wavelength limits a section of the wavefront (a line in the present case) that can be considered as a point source. This source emits approximately semispherical waves that propagate in space; in the case at hand these are approximately semicircles with differences in radii between the neighboring waves equal to one wavelength.

6.31. Since the frequency of the oscillations remains constant, the energy carried by the wave is determined uniquelly by the amplitude, that is, is proportional to the square of the amplitude. The amplitude at a crest A_1 is equal to the sum of the amplitudes of the incident and reflected waves, A_1 and A_2, while the amplitude at a node, A_n, is equal to the difference between A_1 and A_2:

$$A_1 = A_1 + A_2, \quad A_n = A_1 - A_2.$$

Hence, the amplitudes of the incident and reflected waves are

$$A_1 = \frac{A_1 + A_n}{2}, \quad A_2 = \frac{A_1 - A_n}{2}.$$

Hence,

$$\frac{A_2}{A_1} = \frac{A_1 - A_n}{A_1 + A_n} = \frac{A_1/A_n - 1}{A_1/A_n + 1} = \frac{\delta - 1}{\delta + 1}.$$

The ratio of the energy of the reflected wave to that of the incident wave is equal to the ratio of the squares of the amplitudes:

$$\frac{W_2}{W_1} = \left(\frac{\delta - 1}{\delta + 1}\right)^2.$$

Hence, the ratio of the energy that has passed the obstacle to the energy of the waves incident on the obstacle is

$$\frac{W_3}{W_1} = 1 - \left(\frac{\delta - 1}{\delta + 1}\right)^2 = \frac{4\delta}{(\delta + 1)^2}.$$

When the amplitudes are equal ($\delta = 1$) no standing waves are formed and the entire energy passes the obstacle.

In the theory and practice of propagation of waves (say, electromagnetic waves) a common notion is that of the standing-wave ratio, which is the ratio of the energies (or squares of amplitudes) at crest and node. Obviously, in an ideal standing wave this ratio is infinite.

6.32. The figure accompanying the problem shows that the amplitude decreases ten-fold over a distance equal to four wavelengths. Denoting the amplitude near the source by A_0 and the amplitude at a distance of four wavelengths from the source by A_4, we can write

$$A_0/A_4 = 10, \quad \text{or} \quad \log(A_0/A_4) = 1.$$

In natural logarithms,

$$\ln(A_0/A_4) = 2.3.$$

For the amplitude at a distance of one wavelength from the source we have

$$\ln(A_0/A_1) = 2.3/4 = 0.575,$$

while for the amplitude at a distance of z from the source we have

$$\ln(A_0/A_z) = 0.575z/\lambda.$$

Whence

$$A_z = A_0 \exp(-0.575z/\lambda).$$

This dependence is often expressed in terms of the wave number k, which is related to the wavelength as follows: $k = 2\pi/\lambda$. Thus,

$$A_z = A_0 \exp(-0.0916\, kz).$$

6.33. The statement is false. The density of the gas, which is in the denominator of formula (6.33.1), is determined by the ideal-gas law thus:

$$\rho = pM/RT, \qquad (6.33.1)$$

where M is the molecular mass (weight) of the gas, and R is the universal gas constant. If we substitute this value of the density into (6.33.1), the pressure cancels out and we get the formula

$$c = \sqrt{\gamma RT/M}, \qquad (6.33.2)$$

according to which for given gas the speed of sound depends only on the temperature of the gas. Actually, the temperature dependence is somewhat more complicated than simple proportionality to $T^{1/2}$, since in diatomic and especially multiatomic gases the specific heat capacity at constant volume grows noticeably with temperature.

6.34. According to formula (6.33.2), the speed of sound in a gas is proportional to the square root of γ and inversely proportional to the molecular mass. At a fixed temperature the difference in speeds of sound is determined by the ratio γ/M. For water vapor (six degrees of freedom) $\gamma = 1.33$ and for neon (three degrees of freedom) $\gamma = 1.67$. The molecular mass of water is 1.8×10^{-2} kg/mol and that of neon is 2.02×10^{-2} kg/mol. The ratios γ/M is 74.1 for water vapor and 82.5 for neon.

Thus, the upper straight line depicts the temperature dependence of the speed of sound in neon and the lower one depicts the temperature dependence of the speed of sound in water vapor. Both straight lines have the same slope equal to 0.5. A calculation via formula (6.33.2) yields 454 m/s for neon at 300 K and 430 m/s for water vapor at the same temperature.

6.35. When the source is moving and the receiver is fixed, the registered frequency is

$$\nu_1 = \nu_0 \frac{1}{1 - v/c} ,$$

while when the source is fixed and the receiver is moving,

$$\nu_2 = \nu_0 \, (1 + v/c).$$

The first formula implies that ν_1 grows without limit as v/c tends to unity (curve *1* in the figure accompanying the problem), while ν_2 increases linearly as v/c tends to unity (curve *2* in the same figure).

6.36. When the train is moving with a speed v and the speed of sound is c and the frequency measured by an observer on the train is ν_0 (better to say, when the train is at rest), the frequency registered when the train approaches the observer standing at the roadbed is

$$\nu_1 = \frac{\nu_0}{1 - v/c} , \qquad (6.36.1)$$

while the frequency registered when the train is moving away from the observer is

$$\nu_2 = \frac{\nu_0}{1 + v/c} . \qquad (6.36.2)$$

For the sake of brevity we introduce the notation $\nu_1/\nu_2 = \delta$ and $v/c = \beta$. Then

$$\delta = \frac{1 + \beta}{1 - \beta} ,$$

whence $\beta = (\delta - 1)/(\delta + 1)$, or

$$v = \frac{\nu_1 - \nu_2}{\nu_1 + \nu_2} \, c. \qquad (6.36.3)$$

Substituting (6.36.3) into (6.36.1) or (6.36.2), we get

$$\nu_0 = \nu_1 \, (1 - v/c) = \nu_2 \, (1 + v/c) = \frac{2\nu_1 \nu_2}{\nu_1 + \nu_2} .$$

6.37. When the observer stands far from the line along which the source of sound is moving, the equation that describes the Doppler effect contains not the velocity of the sound proper but its projection on the direction of propagation of the wave. For the observer that stands very near to the moving train this velocity is practically that of the train and varies suddenly, and so does the pitch of the sound heard by that observer (curve *1* in the figure accompanying the problem). For the observer that stands at a rather big distance from the moving train, the projection of the velocity varies more smothly, dropping to zero when the train is closest to that observer and then increasing. For this reason the time it takes the registered frequency to change is greater (curve *2*).

6.38. If for an immobile source the wavelength is λ_0, the wavelength λ when the source moves with a velocity v is shorter than λ_0 by vT_0. The waves will arrive at the obstacle having the frequency

$$\nu_1 = \frac{c}{\lambda} = \frac{c}{\lambda_0 - vT} = \nu_0 \frac{1}{1 - v/c}.$$

The waves will reflect from the obstacle but will retain their frequency and wavelength. Since the receiver is moving toward the waves with a velocity v with respect to the medium, the relative velocity of the receiver and waves is $c + v$ and the registered frequency is

$$\nu_2 = \frac{c+v}{\lambda} = \frac{c+v}{c/\nu_0 - v/\nu_0} = \nu_0 \frac{c+v}{c-v} = \nu_0 \frac{1+v/c}{1-v/c}.$$

6.39. At frequency ν_0 the wavelength in still water is $\lambda_0 = c/\nu_0$. In a river whose waters flow with a velocity v, the wavelength downstream is by vT longer than λ_0 and the wavelength upstream is by vT shorter, that is,

$$\lambda = \lambda_0 \pm vT.$$

In relation to the receiver that is down the stream, the velocity of the received waves is the sum of the velocity of waves in still water and the velocity of the river waters (as if the receiver was moving against the waves). For the receiver that is up the stream the velocities are subtracted from each other, with the result that

$$c = c_0 \pm v.$$

The frequency ν registered by a receiver is the ratio of the speed of sound to the wavelength, or

$$\nu = \frac{c_0 \pm v}{\lambda_0 \pm vT} = \frac{c_0 \pm v}{c_0/\nu_0 \pm v/\nu_0} = \nu_0.$$

We see that ν is equal to the frequency of the oscillations generated by the source.

5.40. The wavelength of waves generated by a source moving in a stationary medium is

$$\lambda = \lambda_0 \pm vT,$$

where the minus sign corresponds to the propagation of waves from the source forward, while the plus sign corresponds to waves propagating backward. When the receiver is in motion, its velocity with respect to the waves is

$$c = c_0 \pm v.$$

Here the plus corresponds to motion against the waves, while the minus corresponds to motion in the same direction as the waves propagate. Since the velocities of the boats in relation to waves are different and the distance between the boats remains unchanged, the time it takes a signal to travel from one boat to the other depends on which boat is the receiver and which boat is the source:

$$t = \frac{l}{c_0 \pm v}.$$

If the boats could move with a speed equal to the speed of waves, then the boat moving ahead of the other one would cease to receive any signal, since the signal could not catch up with it. The frequency of the signal received by each boat is defined as the ratio of the velocity with respect to the waves to the receiver wavelength. For the boat floating at the rear,

$$\nu = \frac{c_0 + v}{\lambda_0 + vT} = \frac{1 + v/c_0}{(1 + v/c)\, \nu_0^{-1}} = \nu_0,$$

and for the boat floating in front,

$$\nu = \frac{c_0 - v}{\lambda_0 - vT} = \frac{1 - v/c_0}{(1 - v/c)\, \nu_0^{-1}} = \nu_0.$$

In both cases the frequency of the received signal is equal to that of the sent signal.

6.41. The times of arrival of longitudinal and transverse waves are, respectively,

$$t_{\parallel} = S/v_{\parallel} \text{ and } t_{\perp} = S/v_{\perp},$$

where $v_{\parallel}$ and $v_{\perp}$ are the velocities of propagation of the longitudinal and transverse waves, and S is the distance

272

between A and B. The time interval between the arrival of longitudinal and transverse waves is

$$\Delta t = t_\perp - t_\parallel = S \left(\frac{1}{v_\perp} - \frac{1}{v_\parallel} \right),$$

whence

$$S = \frac{v_\parallel v_\perp}{v_\parallel - v_\perp} \Delta t.$$

If the seismographs are placed at two points, then by measuring the distances S_1 and S_2 (see the figure accompanying the answer) we can establish at which point the source of explosion is located. In fact, in this way the epicenters of earthquakes are located.

6.42. The speed of sound waves in air is $c_1 \approx 330$ m/s and in water it is $c_2 \approx 1500$ m/s. According to Snell's law,

$$\sin \alpha_1 / \sin \alpha_2 = c_1/c_2.$$

Fig. 6.41

Accordingly, when the "sound beam" enters the water, it will be deflected from the perpendicular line still stronger and angle α_2 becomes greater than angle α_1. The velocity ratio determines the maximal angle at which sound waves can go "into" water. The maximal angle of incidence α_m satisfies the condition $(\alpha_2 = 90°)$

$$\sin \alpha_m = c_1/c_2.$$

At $c_1 = 330$ m/s and $c_2 = 1500$ m/s we have $\sin \alpha_m = 0.22$ and $\alpha_m \approx 13°$. At an angle greater than $13°$ total reflection occurs. Such a situation is depicted in the figure accompanying the problem.

The perturbation caused by the incident wave penetrates the surface of the water but dies out exponentially, and this happens the faster the greater the angle of incidence of the wave. The wave dies out practically at a depth of the order of one wavelength. Sometimes one can hear a fisherman whisper: "Keep quiet! The fish is here!" The above estimate shows that a person standing at a distance away from the riverbank can never "scare" the fish.

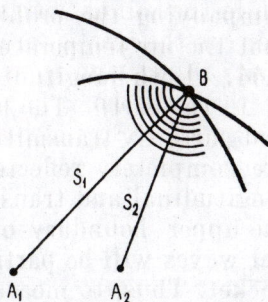

18–01569

6.43. Imagine a plane that is parallel to the surface of the earth. The sound that an explosion generates and that propagates at a certain angle α to the normal to this plane will be deflected still greater. As Snell's law shows, this happens when the speed of sound increases with altitude. Thus, the curve that represents the path along which the sound wave propagates suggests that the speed of sound increases continuously with altitude. Since the speed of propagation of waves in a gas is proportional to the square root of the temperature, then, hence, the behavior of the curve of sound propagation (see the figure accompanying the problem) can be explained by the fact that the air temperature increases with altitude.

6.44. Both longitudinal and transverse waves can travel in the earth. The first are partially reflected by water and partially transmitted through water, while the second are completely reflected by water. The reflection of the longitudinal and transverse waves can be used to estimate the upper boundary of the water pocket. The longitudinal waves will be partially reflected by the bottom of the pocket. Thus, to measure the depth of the pocket one can use only longitudinal waves.

6.45. For the observer to hear the sound of the airplane from a distance a earlier than the sound arrives from point A that is directly above the observer, the time it takes the sound to travel from airplane to observer must be shorter than the time it takes the airplane to fly the distance a plus the time it takes the sound to travel from point A to the observer. The first time is

$$t_1 = \sqrt{a^2 + h^2}/c,$$

while the second is

$$t_2 = a/v + h/c,$$

where c is the speed of sound. The above-stated condition can be written thus:

$$\frac{\sqrt{a^2 + h^2}}{c} < \frac{a}{v} + \frac{h}{c}.$$

If we square both sides of this inequality and carry out the necessary manipulations, we get

$$a\left(\frac{v^2}{c^2} - 1\right) < \frac{2v}{c}\, h. \qquad (6.45.1)$$

The ratio $v/c = M$ is known as the Mach number. Then (6.45.1) can be written thus:

$$a < 2\,\frac{M}{M^2-1}\,h.$$

If, say, the airplane is flying with a speed double the speed of sound, the maximal distance from which the sound will arrive sooner than when the airplane appears overhead is equal to $(4/3)h$.

7. Alternating Current

7.1. The segment of the cross section of the loop of width dr and height h is penetrated by a magnetic flux whose instantaneous value is

$$d\Phi = Bh\,dr,$$

where $B = \mu_0 I/2\pi r$. Whence

$$\Phi = \frac{\mu_0 hI}{2\pi}\int_{R_1}^{R_2}\frac{dr}{r} = \frac{\mu_0 hI}{2\pi}\ln\frac{R_2}{R_1}.$$

The flux coupled with the loop is

$$\Psi = \frac{\mu_0 hIN}{2\pi}\ln\frac{R_2}{R_1}.$$

Fig. 7.2

The current in the conductor is $I = I_0\cos\omega t$. The emf induced in the loop is

$$\mathscr{E}_i = -\frac{d\Psi}{dt} = \frac{\mu_0 hI_0 N\omega}{2\pi}\ln\frac{R_2}{R_1}\sin\omega t.$$

Finally, the effective value of this emf is

$$\mathscr{E}_{i\,\mathrm{eff}} = \frac{\mu_0 hN\omega I_{\mathrm{eff}}}{2\pi}\ln\frac{R_2}{R_1}.$$

7.2. The figure accompanying the problem shows that the capacitive reactance is four times the inductive reactance. If the frequency is doubled, the first quantity will decrease by a half and the second will double, which means they will become equal. As shown by the figure accompanying the answer, the ratio $\mathscr{E}_0/I_0$ will decrease,

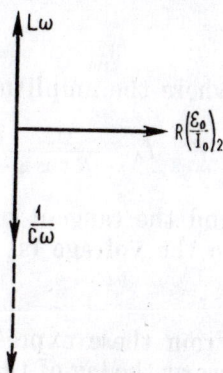

and since $\mathscr{E}_0$ must remain unchanged, the current grows. The same result can be obtained analytically. The amplitude of the current in the circuit is

$$I_0 = \frac{\mathscr{E}_0}{\sqrt{R^2 + \left(\dfrac{1}{C\omega} - L\omega\right)^2}}.$$

Prior to the change in frequency, $1/C\omega > L\omega$, and hence

$$\left(\frac{1}{C\omega} - L\omega\right)^2 > 0.$$

After the frequency is doubled, $1/C\omega = L\omega$. Here $I_0 = \mathscr{E}_0/R$.

7.3. The current in the circuit containing a resistance and an inductance connected in series is

$$I = I_0 \sin(\omega t + \varphi),$$

where the amplitude value of the current is

$$I_0 = \frac{\mathscr{E}_0}{R\cos\varphi - L\omega\sin\varphi}, \quad \text{or} \quad I_0 = \frac{\mathscr{E}_0}{\sqrt{R^2 + L^2\omega^2}},$$

and the tangent of the phase of the current with respect to the voltage is

$$\tan\varphi = -L\omega/R.$$

From these expressions it follows that as the frequency grows the lag of the current phase in relation to the voltage phase increases, which results in a decrease in the current. The average power in the circuit is defined thus:

$$P = \frac{1}{2}\mathscr{E}_0 I_0 \cos\varphi.$$

As the frequency grows, the amplitude of the current decreases and so does the power factor, which is the cosine of the phase shift between voltage and current. The power will also decrease as a result.

7.4. The current in the circuit containing a resistance and a capacitance connected in series is

$$I = I_0 \sin(\omega t + \varphi),$$

where the amplitude value of the current is

$$I_0 = \frac{\mathscr{E}_0}{R\cos\varphi + (1/C\omega)\sin\varphi}, \quad \text{or} \quad I_0 = \frac{\mathscr{E}_0}{\sqrt{R^2 + 1/C^2\omega^2}},$$

and the tangent of the phase shift of the current with respect to the voltage is

$$\tan \varphi = 1/RC\omega.$$

From these expressions it follows that as the frequency grows the phase shift by which the current leads the voltage decreases and tends to zero, while the current grows. The average power in the circuit, defined as

$$P = \frac{1}{2}\, \mathscr{E}_0 I_0 \cos \varphi,$$

increases with frequency, since $\cos \varphi$ tends to unity, and so does the amplitude value of the current.

7.5. The figure accompanying the problem shows that the current leads the voltage in the phase by $0 < \varphi < \pi/2$. This happens if a capacitance is connected in series with the resistance.

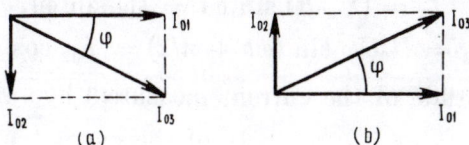

Fig. 7.6

7.6. For the case shown in Figure (a) accompanying the problem, we can write (if we ignore the resistances of the ammeters)

$$I_1 = \frac{U_0}{R} \sin \omega t = I_{01} \sin \omega t,$$

$$I_2 = \frac{U_0}{L\omega} \sin \left(\omega t - \frac{\pi}{2} \right) = -I_{02} \cos \omega t,$$

where U_0 is the amplitude value of the voltage between points *1* and *2*. The current I_3 flowing through ammeter *A3* is the sum of currents I_1 and I_2:

$$I_3 = I_{03} \sin (\omega t + \varphi)$$

(the vector diagram of currents is depicted in Figure (a) accompanying the answer). The amplitude value of current I_3 is

$$I_{03} = \sqrt{I_{01}^2 + I_{02}^2},$$

and the phase shift of the current in relation to the voltage is

$$\tan \varphi = -I_{02}/I_{01}.$$

Since the ammeters measure the effective value of the current, $I_{\text{eff}} = I_0/\sqrt{2}$, we have

$$I_{3\text{eff}} = \sqrt{I_{1\text{eff}}^2 + I_{2\text{eff}}^2}.$$

In the case shown in Figure (b) accompanying the problem, just like in the previous one, the currents that flow through the resistance and the capacitance differ in phase by $\pi/2$, the only difference being that here the current flowing through the capacitance leads the applied voltage, while the current flowing through the inductance lags behind the voltage. The corresponding vector diagram is depicted in Figure (b) accompanying the answer. The currents measured by ammeters $A1$ and $A2$ are

$$I_1 = (U_0/R) \sin \omega t = I_{01} \sin \omega t,$$
$$I_2 = U_0 C \omega \sin (\omega t + \pi/2) = I_{02} \cos \omega t.$$

The amplitude of the current measured by ammeter $A3$ is

$$I_{03} = \sqrt{I_{01}^2 + I_{02}^2},$$

while the tangent of the phase shift is

$$\tan \varphi = I_{02}/I_{01}.$$

The current measured by ammeter $A3$ is

$$I_{3\text{eff}} = \sqrt{I_{1\text{eff}}^2 + I_{2\text{eff}}^2} < I_{1\text{eff}} + I_{2\text{eff}}.$$

7.7. For the case depicted in Figure (a) accompanying the problem, the voltage between points 1 and 2 is

$$U_1 = I_0 R \sin \omega t = U_{01} \sin \omega t,$$

while that between points 2 and 3 is

$$U_2 = \frac{I_0}{C\omega} \sin \left(\omega t - \frac{\pi}{2} \right) = -\frac{I_0}{C\omega} \cos \omega t = -U_{02} \cos \omega t$$

(see the vector diagram in Figure (a) accompanying the answer). The voltage between points 1 and 3 is the sum of U_1 and U_2:

$$U_3 = U_1 + U_2 = U_{03} \sin (\omega t + \varphi).$$

Its amplitude value is

$$U_{03} = \sqrt{U_{01}^2 + U_{02}^2},$$

while the phase shift with respect to the applied voltage is given by the following formula:

$$\tan \varphi = -U_{02}/U_{01}.$$

Since the voltmeters measure the effective value $U_{eff} = U_0/\sqrt{2}$, we have

$$U_{3eff} = \sqrt{U_{1eff}^2 + U_{2eff}^2} < U_{1eff} + U_{2eff}.$$

For the circuit depicted in Figure (b) accompanying the problem, just like in the previous case, the voltages

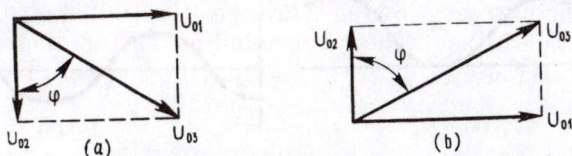

Fig. 7.7

across the resistance and the inductance differ in phase by $\pi/2$, the only difference being that here the current flowing through the inductance lags behind the voltage, while in the previous case the current flowing through the capacitance leads the voltage (and, hence, the phase shift between the voltages across the resistance and across the capacitance is $-\pi/2$). The respective voltages are

$$U_1 = I_0 R \sin \omega t = U_{01} \sin \omega t,$$
$$U_2 = I_0 L\omega \sin (\omega t + \pi/2) = U_{02} \cos \omega t$$

and

$$U_3 = U_3 \sin (\omega t + \varphi)$$

(see the vector diagram depicted in Figure (b) accompanying the answer). The amplitude value of the voltage is

$$U_{03} = \sqrt{U_{01}^2 + U_{02}^2}.$$

The effective voltages measured by the voltmeters are related thus:

$$U_{3eff} = \sqrt{U_{1eff}^2 + U_{2eff}^2} < U_{1eff} + U_{2eff}.$$

The tangent of the phase of the voltages is

$$\tan \varphi = U_{02}/U_{01}.$$

7.8. In the first case we have resonance, at which the voltages across the capacitor and the inductance,

$$U_C = \frac{I_0}{C\omega} \sin(\omega t - \pi/2) \text{ and }$$

$$U_L = I_0 L\omega \sin(\omega t + \pi/2), \quad (7.8.1)$$

are equal in magnitude and opposite in phase. From Eq. (7.8.1) and the fact that a capacitor and an inductance

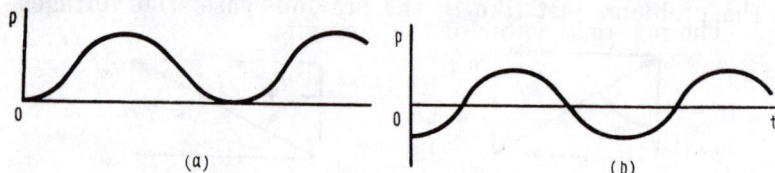

Fig. 7.9

connected in series do not change the current it follows that

$$1/C\omega = L\omega.$$

For the case where a capacitance and an inductance are connected in parallel, in each of these elements there flows a current

$$I_C = U_0 C\omega \sin(\omega t - \pi/2) \text{ and } I_L = \frac{U_0}{\omega L} \sin(\omega t + \pi/2).$$

The total current is

$$I = U_0 (C\omega - 1/L\omega) \cos \omega t,$$

and, since $C\omega = 1/L\omega$, we have

$$I = I_C + I_L = 0.$$

7.9. If the voltage varies according to the law

$$U = U_0 \sin \omega t$$

and there is a definite phase shift between voltage and current, so that

$$I = I_0 \sin(\omega t + \varphi)$$

(where the phase difference φ may be either positive or negative), then the instantaneous value of the power is

$$P = U_0 I_0 \sin(\omega t + \varphi) \sin \omega t.$$

If we write

$$\sin(\omega t + \varphi) \sin \omega t = \sin^2 \omega t \times \cos \varphi + \sin \omega t \times \cos \omega t \times \cos \varphi$$

$$= \frac{1}{2}[(1 - \cos 2\omega t) \cos \varphi + \sin 2\omega t \times \sin \varphi],$$

we get

$$P = \frac{1}{2}[\cos \varphi - \cos(2\omega t + \varphi)] U_0 I_0. \qquad (7.9.1)$$

The maximal value of the power is

$$P_{\max} = \frac{1}{2} U_0 I_0 (\cos \varphi + 1),$$

while the minimal value is

$$P_{\min} = \frac{1}{2} U_0 I_0 (\cos \varphi - 1).$$

Whence, the power factor is

$$\cos \varphi = \frac{P_{\max} + P_{\min}}{P_{\max} - P_{\min}}$$

(bear in mind that $P_{\min}$ is negative).

Formula (7.9.1) shows that the frequency of power variation is twice the frequency of the applied voltage. During one period of voltage variation the power passes twice through the maximum and the minimum.

Here are some particular cases.

(1) $\varphi = 0$. The load is a purely active resistance. In this case (Figure (a) accompanying the answer) $P_{\min} = 0$ and $P_{\max} = U_0 I_0$.

(2) $\varphi = \pm \pi/2$. The circuit contains only a reactive element, that is, a capacitance or an inductance. Since in this case $\cos \varphi = 0$, we have (see Figure (b) accompanying the answer)

$$P_{\max} = -P_{\min}.$$

The work performed by the AC source over one period of variation of the power is zero. This means that during one half of the period the energy flows from the AC source to the reactive element in the form of the electrostatic energy of the capacitance or the magnetic energy of the inductance, while during the other half the energy is returned to the AC source.

7.10. When a watch is inside the solenoid, the magnetic field generated by the solenoid forces the steel parts of

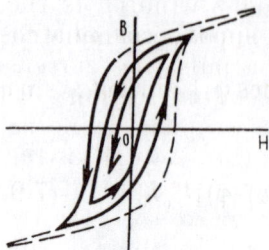

Fig. 7.10

the watch to change periodically their magnetization, following the hysteresis loop. When the watch is slowly removed from the solenoid, the magnetic field acting on the watch gradually decreases, and as the periods pass, the hysteresis loop shrinks. Each second 50 hysteresis loops are traversed, each being smaller than the previous one (the number "50" appears because the frequency of the AC source is usually 50 Hz). This process is roughly sketched in the figure. When the watch is completely removed from the magnetic field, it proves to be completely demagnetized.

7.11. At the moment when the "plus" of the voltage is at terminal a (see the figure accompanying the problem),

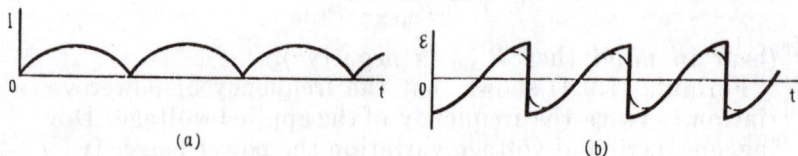

Fig. 7.11

the current passes through diode 2, resistor R, diode 3, and returns to the AC source through terminal b, which has the "minus" of the voltage at that moment. After the applied voltage changes sign, the current from terminal b passes through diode 4, resistor R, diode 1, and returns to the AC source via the negative terminal a. Thus, the current that passes through R consists of a series of alternating halves of sinusoids (Figure (a)). The average value of the current over one or any integral number of half-periods is

$$I_{av} = \frac{2I_0}{T\omega} \int_0^{T/2} \sin \omega t \, dt = \frac{4I_0}{T\omega} = \frac{2I_0}{\pi} \approx 0.637 I_0.$$

282

In carrying out this calculation in accordance with the conditions of the problem, it was assumed that the voltage drop across the diodes is negligible and that the rectification process does not alter the sinusoidal nature of the emf. As for the emf that is generated in the secondary winding of a transformer whose primary winding is the load resistance R, it must have two opposite symmetric sections, since each half-period of the pulsating current has an ascending section and a descending section. An idealized curve of the voltage in the secondary winding of a transformer is shown in Figure (b). Actually, the curve is much smoother because of the inductance of the transformer, which plays the role of a choke coil, the interturn capacitance, and other factors. The approximate shape of the voltage curve on the transient sections is depicted by a dashed curve.

7.12. After the rectifier the current branches out (see the arrows in the figure accompanying the problem). A

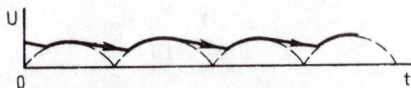

Fig. 7.12

fraction of the current flows through resistor R and a fraction is used to charge the capacitor. If the internal resistance of the source (together with the diodes) is low, then the voltage across the capacitor is equal to the voltage at the "out" terminals. This occurs as long as the voltage is lower than the maximum of the pulsating voltage. After the voltage passes the maximum, it falls off and becomes lower than the voltage across the capacitor. Because of this the capacitor will begin to discharge through the resistor, with the voltage across the capacitor decaying according to the law

$$U = U_0 \exp\left(-t/RC\right)$$

(the discharge current is designated by arrows in the figure accompanying the answer). The greater the capacitance, the slower the decay, which continues until the voltage across the capacitor becomes equal to the growing voltage in the following half-wave. Then the capacitor is charged to the maximum of the voltage anew. The process continues in this manner. Thus, a capacitor in the circuit makes the "out" voltage smoother, and the higher the capacitance the stronger the effect. The curve representing the time variation of the current flowing through the resistor follows the voltage curve in parallel.

7.13. For both directions of the emf applied to the transformer, the current is limited by the diode introduced into the circuit in the blocking direction. This current is

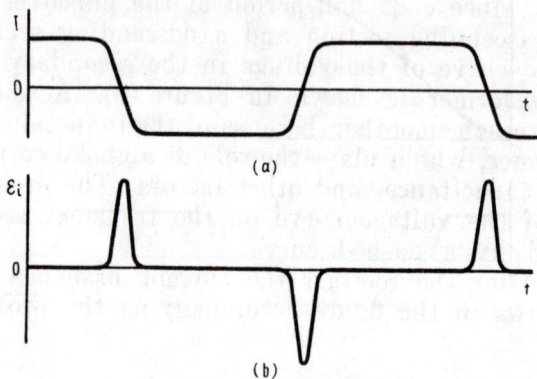

Fig. 7.13

caused by the motion of the minority (intrinsic) charge carriers and reaches a plateau very rapidly as the voltage is increased. The diode introduced in the conducting direction does not limit the current. For this reason, the oscillogram of the current in the primary circuit has the form shown in Figure (a). Accordingly, the greater fraction of time in each half-period (in each direction) the emf induced in the secondary winding is zero. Only over small time intervals when the current passes through zero does an emf emerge, first in one direction and then in the other (Figure (b)). The oscillograms here are, of course, only rough sketches, since they do not take into account the inductances in the transformer circuits. Note that in modern semiconductor diodes the reverse current is negligible, with the result that the problem is of purely academic interest.

7.14. In some respects this problem resembles the previous one. Here, too, the current in the primary circuit is limited to the saturation current in one of the diodes, introduced into the circuit in the conducting direction rather than in the blocking. In contrast to Problem 7.13, the present one possesses a special feature that manifests itself in the initial section near the zero of the current in the circuit. While in a semiconductor diode the current increases with voltage almost linearly in the initial section, in a vacuum diode the voltage dependence of the

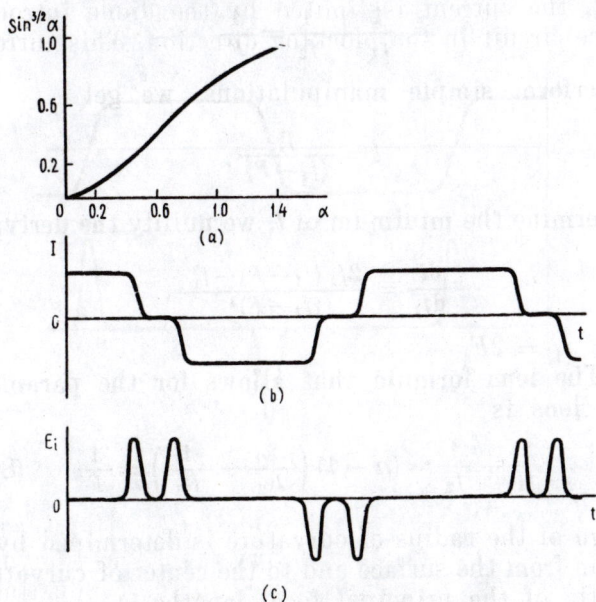

Fig. 7.14

current is described with sufficient accuracy by the three-halves power law $I = KU_0^{3/2}$.

The constant K incorporates universal constants and the distance between the electrodes in the diode. Since the voltage varies with time according to the sinusoidal law, the current flowing through the diode on the initial section of the voltage increase must be written in the form

$$I = KU_0^{3/2} \sin^{3/2} \omega t$$

(the function $f(\alpha) = \sin^{3/2} \alpha$ is depicted in Figure (a)). Allowing for this dependence, we obtain the oscillograms of current in the primary circuit (Figure (b)) and of the emf in the secondary circuit (Figure (c)). Just as in the previous problem, we have not allowed for the effects associated with the presence of inductances in the transformer circuits.

8. Optics

8.1. If we introduce the notation $l = f_1 + f_2$ in the lens formula

$$\frac{1}{f_1} + \frac{1}{f_2} = \frac{1}{F}$$

and perform simple manipulations, we get

$$l = \frac{f_1^2}{(f_1 - F)} .$$

To determine the minimum of l, we nullify the derivative

$$\frac{dl}{df_1} = \frac{2f_1(f_1 - F) - f_1^2}{(f_1 - F)^2} ,$$

whence $f_1 = 2F$.

8.2. The lens formula that allows for the parameters of the lens is

$$\frac{1}{f_1} + \frac{1}{f_2} = (n-1)\left(\frac{1}{R_2} - \frac{1}{R_1}\right) = \frac{1}{F} \qquad (8.2.1)$$

(the sign of the radius of curvature is determined by the direction from the surface and to the center of curvature). The ratio of the principal focal lengths is

$$\frac{F_a}{F_b} = \frac{n_b - 1}{n_a - 1} , \qquad (8.2.2)$$

where we have allowed for the fact that the radii of curvature of both lenses are the same. Formula (8.2.1) can be transformed thus:

$$f_2 = \frac{f_1 F}{f_1 - F} .$$

On the curve representing the f_2 vs. f_1 dependence, the value of F is determined by the position of the vertical asymptote of each curve. However, a more exact value can be obtained by drawing a straight line that passes through the origin at an angle of $45°$ to the axes. In this case the coordinates of the points of intersection of this straight line with the curves yield $f_2 = f_1 = 2F$ for both lenses, while the ratios of these coordinates determine, via formula (8.2.2), the ratio of $n_b - 1$ to $n_a - 1$.

8.3. The smaller the aperture, the lower the optical

distortions caused by the large width of the beam of light incident on the lenses of the objective. If the aperture is very small, the optical properties of the camera closely resemble those of a pinhole camera, whose aperture, in terms of geometrical optics, can be as small as desired and

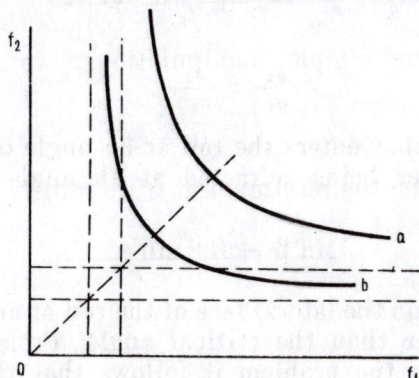

Fig. 8.2

whose depth of focus extends from zero to infinity. Actually, however, diffraction imposes certain restrictions on this ideal case. The limiting value of the diameter of the aperture, D, is determined by the wavelength of the light and by the distance from the aperture to the photographic plate. Theoretical considerations suggest that D must be close to the value for which only one Fresnel zone fits into the aperture:*

$$D \approx 4\sqrt{\lambda f}.$$

For instance, at $\lambda \approx 0.5$ μm and $f \approx 5$ cm, the diameter of the aperture is approximately 0.6 mm. Note that in photography the size of the aperture is characterized by a quantity known as the aperture ratio, or the ratio of the diameter of the aperture to the focal length. Usually the aperture ratio is marked by a fraction whose numerator is unity (1/4.5, 1/5.6, 1/8, 1/11). In the example we are discussing here the aperture ratio is equal to 1/80. In cameras the smallest aperture ratio is practically never less than 1/16, so that diffraction effects play no role in the present problem and need not be taken into account.

* According to Rayleigh, the sharpest focus in a pinhole camera is achieved when the radius of the aperture is 0.95 of the radius of the zeroth Fresnel zone.

8.4. The solution can be found from simple trigonometric reasoning under common assumptions and approximations:

$$\sin \alpha_1 \approx \tan \alpha_1 \approx \alpha_1, \quad \sin \alpha_2 \approx \tan \alpha_2 \approx \alpha_2,$$

$$\alpha_1/\alpha_2 \approx n, \quad y_1/a_1 \approx \alpha_1, \quad y_2/a_2 \approx \alpha_2.$$

Whence,

$$\frac{y_2}{y_1} = \frac{a_2}{a_1 n}.$$

8.5. A ray that enters the rod at an angle α, travels in the glass after being refracted at an angle β given by Snell's law:

$$\sin \beta = n^{-1} \sin \alpha. \tag{8.5.1}$$

The ray falls on the lateral face of the rod at an angle that is not smaller than the critical angle. From the figure accompanying the problem it follows that this angle is $\pi/2 - \beta$. According to the critical angle condition,

$$\sin (\pi/2 - \beta) = \cos \beta \geqslant n^{-1}. \tag{8.5.2}$$

The maximal value of β at $\alpha = \pi/2$, according to (8.5.1), obeys the condition

$$\sin \beta = 1/n. \tag{8.5.3}$$

Squaring (8.5.2) and (8.5.3) and adding the squares, we get

$$1 \geqslant 2/n^2,$$

whence

$$n \geqslant \sqrt{2}.$$

The phenomenon of light "trapping" in a glass rod is widely used in fiber optics. If the attenuation of light in the glass is low, the ray can travel over great distances. Bundles of such rods (or fibers) form cables over which data can be transmitted with a high accuracy and a low level of noise. Internal organs of human beings can be illuminated with the light transmitted by such fibers, which at present is widely used in medical practice for diagnostic purposes.

8.6. The figure accompanying the problem shows that after reflection from the first mirror the beam changes its direction by an angle of 2α, while after reflection from the second mirror the beam changes its direction by an additional angle of 2β. For the refracted beam to travel in the direction opposite to the direction of the incident beam, the sum $2\alpha + 2\beta$ must be equal to π, or $\alpha + \beta = \pi/2$.

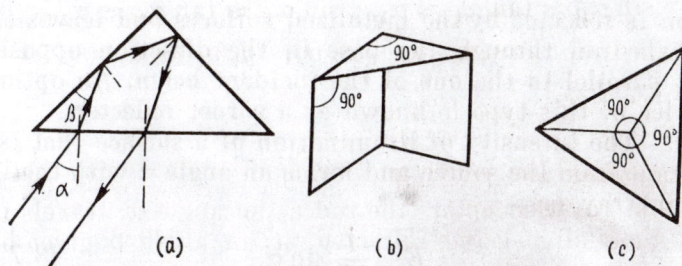

Fig. 8.6

In this case the angle between the normals to the mirrors is

$$\theta = \pi - (\alpha + \beta) = \pi/2.$$

The angle between the mirrors must also be equal to $\pi/2$.

If instead of the two mirrors we take a prism (see Figure (a) accompanying the answer), then a beam incident on the base of the prism at an angle α will enter the prism at an angle β determined by Snell's law. For the refracted beam to leave the prism in the direction opposite to the one of the incident beam after undergoing total internal reflection from the lateral surfaces of the prism, the beam must fall on the base of the prism (after it has been reflected by the second lateral surface) at an angle β. Figure (a) accompanying the answer shows that the beam travels the same path as in the case of two mirrors, whereby the angle at the apex of the prism must be equal to $\pi/2$. We see that a prism may also be used to reverse a beam. For the beam to retain its energy after traveling through the prism practically for all angles of incidence, the lateral surfaces of the prism must be metalized. If three flat mirrors are positioned at right angles, as shown in Figure (b) accompanying the answer, it can be demonstrated that the beam of light may be oriented with respect to the first mirror (on which it is incident) in an arbitrary manner and yet the refracted beam will always be parallel to the incident one. Instead of three mirrors we can use a glass tetrahedron with a right trihedral angle at the apex and identical metalized lateral surfaces in the form of right isosceles triangles (see Figure (c) accompanying the answer). A beam incident on the base of the tetrahedron is reflected by the metalized surfaces and leaves the tetrahedron through the base in the direction opposite but parallel to the one of the incident beam. An optical device of this type is known as a corner reflector.

8.7. The intensity of illumination of a surface that is r distant from the source and forms an angle α with the incident ray is

$$E = \frac{I}{r^2} \sin \alpha, \qquad (8.7.1)$$

where I is the intensity of the source. At the edge of the table, according to (8.7.1),

$$E = \frac{Ih}{(R^2+h^2)^{3/2}} \cdot$$

To find the maximum of E we must nullify the derivative:

$$\frac{dE}{dh} = I \frac{(R^2+h^2)^{3/2} - 3h^2 (R^2+h^2)^{1/2}}{(R^2+h^2)^3} = 0,$$

whence

$$h = R/\sqrt{2}.$$

8.8. The ratio of the sines of the angles is equal to the ratio of the speeds of light in the media:

$$\sin \alpha_1/\sin \alpha_2 = c_1/c_2.$$

. e ratio of the wavelengths is equal to the ratio of the speeds of light:

$$\lambda_1/\lambda_2 = c_1/c_2.$$

Therefore

$$\lambda_2 = \frac{\sin \alpha_2}{\sin \alpha_1} \lambda_1.$$

8.9. The optical path difference, which determines the interference pattern, is $|z_2 - z_1|/\lambda$. Since $|z_2 - z_1|$ cannot be greater than a, the maximal possible number of fringes on each side from the middle of the screen (i.e. for $z_2 > z_1$ and for $z_2 < z_1$) is equal to the ratio a/λ, while the total number of fringes is $2a/\lambda$. Actually the number of fringes that can be observed is considerably lower, since at $z_2 - z_1 = a$ the interference fringes must lie in the plane in which the sources lie.

8.10. As the source of light is positioned symmetrically in relation to the mirrors, its virtual images appear at equal distances d from the source and, as the figure accompanying the answer shows,

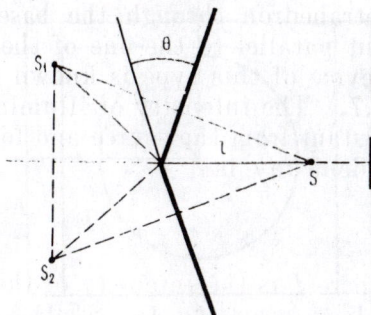

Fig. 8.10

$$d = 2l \cos (\theta/2).$$

The source and its virtual images lie at the vertices of an isosceles triangle. The distance between the virtual images is

$$a = 2d \sin (\theta/2),$$
$$\text{or} \quad a \approx 2l \sin \theta.$$

The first interference fringes on a screen that is L distant from the mirrors are separated by a distance of

$$h = \lambda L/a,$$

and, hence, the smaller the value of θ, the greater the distance h.

8.11. Since equal phase differences correspond to equal optical path differences, we can write

$$(z_2 - z_1)/\lambda = \text{const}, \quad \text{or} \quad z_2 - z_1 = n\lambda,$$

where n is an integer. A surface whose points possess the property that the difference in the distances from any point to two fixed points (the foci) is a constant, constitutes a hyperboloid. The section of this hyperboloid by any plane containing these sources results in two branches of a hyperbola. The sections of the hyperboloid by planes that are perpendicular to the straight line which passes through the middle of the segment connecting the sources are also branches of hyperbolas. For this reason, the observed interference fringes have the form of hyperbolas.

8.12. When light is reflected from the upper boundary of each film, the phase of the wave changes to the opposite or, as it is usually said, a half-wave is lost. The light that passes through the film is reflected by the substrate, which in one case has a refractive index greater than that of the film and in the other, smaller than that of the film. When

$n_2 > n$, a new change in the phase of the reflected wave to the opposite one occurs, while when $n_1 < n$, the phase of the reflected wave is retained. For this reason, the places on one film where light is observed correspond to the dark places on the other film, and vice versa.

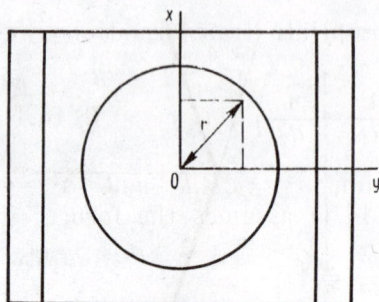

(a)

(b)

(c)

Fig. 8.14

8.13. The difference between neighboring interference fringes in air is determined by the relationship

$$a_0 = \lambda_0/2 \tan \alpha,$$

while for a liquid this relationship is

$$a = \lambda/2 \tan \alpha.$$

Since $\lambda = \lambda_0/n$, we can write $a = a_0/n$.

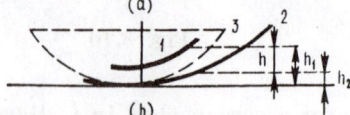

8.14. Interference is caused by the difference in paths of the light rays that forms in the space between the lens and the cylinder. The interference fringes constitute bands of equal width.

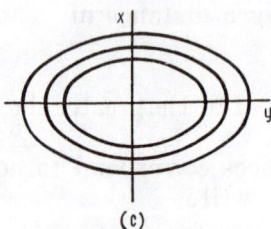

Let us introduce a system of coordinates. One axis, the x axis, is directed along the generator of the cylinder that passes through the point at which the lens touches the cylinder, while the second axis, the y axis, is at right angles to the generator discussed above (see Figure (a) accompanying the answer). We draw a plane that is perpendicular to the x axis and passes at a distance y from the origin. Figure (b) shows the section of the lens by the plane (curve 1) and the section of the cylinder by the plane (curve 2). The same figure demonstrates the section of the lens by a plane that is perpendicular to the x axis and intersects the lens along its diameter (dashed curve 3). From Figure (b) it also follows that the gap between the lens and the cylinder is

$$h = h_1 - h_2 = \frac{r^2}{2R_1} - \frac{y^2}{2R_2} = \frac{x^2 + y^2}{2R_1} - \frac{y^2}{2R_2}.$$

292

Here, as usual, we assume the following approximations to be valid:

$$r = \sqrt{2R_1 h}, \quad y = \sqrt{2R_2 h}.$$

After carrying out the appropriate transformations we get

$$\frac{x^2}{2R_1 h} + \frac{y^2}{2h}\left(\frac{1}{R_1} - \frac{1}{R_2}\right) = 1. \qquad (8.14.1)$$

If we introduce the notation $a^2 = 2R_1 h$ and $b^2 = 2hR_1R_2/(R_2 - R_1)$, then (8.14.1) assumes the form

$$\frac{x^2}{a^2} + \frac{y^2}{b^2} = 1.$$

The interference fringes have the shape of ellipses (see Figure (c) accompanying the answer) in which h is a parameter. In reflected light, $h = (2k + 1)/\lambda$ (with $k = 0, 1, 2, 3, \ldots$) for bright bands and $h = k\lambda$ for dark.

8.15. The section of the cylinder segment by a plane parallel to the plane of the drawing is everywhere the same. For this reason, all points that have the same path difference for the ray reflected from the lower surface of the cylinder and the ray reflected from the upper surface of the plate lie at the same distance from the cylinder's generator that touches the plate, with the result that the interference fringes are in the form of straight lines parallel to the generator. The method of determining the distances between the sequential fringes closely resembles the method of determining the radii of Newton rings. The distances from the generator that touches the plate satisfy the same conditions as the radii of Newton rings do, namely,

$$h = \sqrt{R\lambda\,\frac{k+1}{2}}$$

for bright bands in reflected light and dark bands in transmitted light, and

$$h = \sqrt{R\lambda k}$$

for dark bands in reflected light and bright bands in transmitted light. As we move away from the generator, the distances between neighboring bands become smaller, just as the radii of Newton rings do,

8.16. The width of the air gap between the lens and the plate is the sum of the thickness of the lens section and the particle thickness:

$$h = h_0 + a = \frac{r^2}{2R} + a,$$

where r is the radius of the ring being observed. A bright ring whose number is k is observed at

$$h = \frac{2k+1}{2} \frac{\lambda}{2}.$$

Thus,

$$r^2 = \frac{2k+1}{2} \lambda R - 2Ra.$$

If the numbers of sequential rings are laid off on the horizontal axis and the square of the radii of the corresponding rings, on the vertical axis, we obtain a straight line (see the figure accompanying the answer) whose slope is equal to the ratio of the difference of squares of radii of two neighboring rings to the product λR, that is, $(r_k^2 - r_{k-1}^2)/\lambda R$. Knowing R, we can find λ. Note that in this method there is no difference between bright and dark rings, and knowing the exact number of a ring is not necessary. For this reason, in the figure accompanying the answer we have assigned a number k to an arbitrary ring, while the numbers $k - 1$ and $k + 1$ are assigned to the neighboring rings.

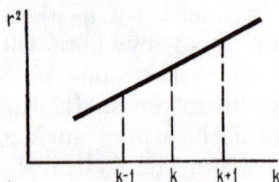

Fig. 8.16

8.17. To construct the interference fringes, we draw a number of straight lines parallel to the plate in such a way that the distances between them along the vertical line are equal to one-half of the wavelength. The points at which these straight lines intersect the substrate (including the surface of the ledge) determine the position of the interference fringes of equal width. Analyzing the position of the fringes obtained here, one can establish that from the wider side of the wedge (in the figure accompanying the problem, on the right) the distance between the fringes, or bands, is smaller (for any value of θ)

than over the flat sections of the substrate. The distance between the fringes from the narrow side of the wedge (on the left) can be either smaller or greater than the distance over the flat sections depending on the relationship between θ and α. For $\theta > \alpha$ (see Figure (a) accompanying the answer),

$$\theta = \alpha + \beta.$$

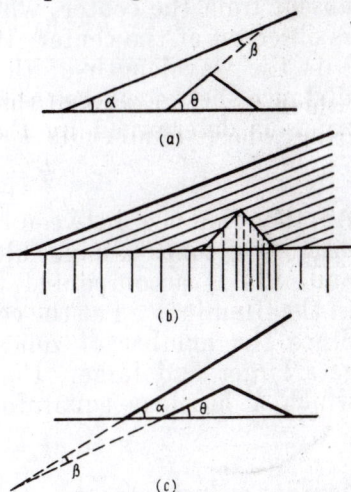

(a)

(b)

(c)

Fig. 8.17

The left side of the ledge acts as a substrate and forms an angle β with the plate, that is, a wedge. If $\theta > 2\alpha$, then $\beta > \alpha$ and the distance between the fringes is smaller than that between the fringes over the flat section of the plate. This case is depicted in Figure (b) and corresponds to the case depicted in the figure accompanying the problem. But if $\theta < 2\alpha$, we have $\beta < \alpha$ and the fringes above the left side of the wedge are separated by a distance greater than that separating the fringes over the flat section of the plate. For $\theta < \alpha$ (see Figure (c) accompanying the answer), the left side of the wedge also acts as a substrate and forms a wedge with an angle $\beta < \alpha$ with the plate. In this case, too, the distance between the fringes is greater than that between the fringes over the flat section of the plate.

8.18. The interference fringes in the wedge constitute bands of equal width. Ledges diminish, while dents increase the width of the air gap where the path difference of rays is formed. For this reason, at the points of a ledge the path difference is the same as at the points of the wedge closer to the narrow part of the gap, while at the points of a dent the difference is the same as at points closer to the wide part of the gap. For this reason, the interference pattern depicted in Figure (b) accompanying the problem corresponds to a ledge, while that depicted in Figure (c) corresponds to a dent.

8.19. The intensity of illumination at the center of the second screen is determined by the number of Fresnel zones into which the section of the wave surface limited by the hole in the first screen can be partitioned. If this number is not large and is even, the light is practically absent from the center, while if the number is odd, light is observed at the center. If a is the diameter of the hole, λ is the wavelength of the incident light, and z is the distance between the screens, the number of Fresnel zones is determined by the expression

$$k = a^2/4\lambda z.$$

As the distance between the screens is increased, the number of zones assumes alternately odd and even values, and this is accompanied by an increase or a decrease in the illumination at the center of the diffraction pattern. Since the number of zones continuously decreases as z gets larger and larger, the limit distance is the one at which k becomes equal to unity, that is,

$$z = a^2/4\lambda.$$

At a distance greater than this value, the intensity decreases monotonically, and for $z \gg a^2/4\lambda$ the intensity changes in inverse proportion to z^2, that is, just like for a point source.

8.20. When the central Fresnel zone and several neighboring zones are screened, the light intensity at the center of the geometric shadow is exactly the same as if one-half of the first nonscreened zones was acting. The calculation is carried out in the same manner as when there is no obstacle, the only difference being that the calculation of the overall action of the Fresnel zones starts not from the zeroth (or central) zone but from the first nonscreened zone. Therefore, a bright spot is always observed at the center of the screen irrespective of the distance to the obstacle or of the wavelength of the light wave (the only requirement is that the number of zones screened by the obstacle be moderate).

A theoretical description of the formation of a bright spot at the center of the geometric shadow was first carried out by Poisson, who used it as an objection against the wave theory of light, since he assumed that such a spot could simply not exist. But an experiment carried out by Arago proved without doubt that such a spot does indeed exist. Actually, this spot was discovered roughly a hundred years earlier by Maraldi. Curiously enough, the spot was later named the Poisson spot.

8.21. The maximum condition in the spectrum of a diffraction grating is

$$c \sin \varphi = k\lambda.$$

Longer wavelengths correspond to larger angles. The figure accompanying the question shows that the position of the second-order maximum of the line λ_2 is close to that of the third-order maximum of the line λ_1. Therefore, $c \sin \varphi = 2\lambda_2 \approx \lambda_3$. Whence, $\lambda_2/\lambda_1 \approx 1.5$.

8.22. The condition for a first-order diffraction maximum to occur is

$$c \sin \varphi_1 = \lambda.$$

For the highest-order maximum we have $c \sin \varphi_m = k_m \lambda$, whence

$$k_m = \frac{\sin \varphi_m}{\sin \varphi_1}.$$

Since the value of $\sin \varphi_m$ cannot exceed unity,

$$k_m \leqslant \frac{1}{\sin \varphi_1}. \qquad (8.22.1)$$

If k_m contains both an integral part and a fractional part, the latter must be discarded irrespective of its value. For instance, if in the first order the line is observed at an angle of 8.36°, formula (8.22.1) yields $k_m \approx 6.88$. The maximal order, therefore, is $k_m = 6$.

8.23. The angles that determine the position of the first maximum for both gratings are the same, which means that the gratings spacings are the same. To estimate the resolving power, we must find the ratio of the wavelength at the maximum of a line to the difference between this wavelength and the wavelength corresponding to a neighboring minimum. For small angles the sine function may be replaced with the angles, so that

$$\varphi_{max} \approx \lambda_{max}, \quad \varphi_{min} \approx \lambda_{min}.$$

The resolving power,

$$\delta = \frac{\lambda}{\lambda_{max} - \lambda_{min}},$$

is equal approximately to 25 for grating *1* and 10 for grating *2*.

8.24. The resolving power of a grating is

$$\delta = kN, \qquad (8.24.1)$$

where N is the general number of lines (or grooves), and k is the order of the spectrum. The maximal resolving power is determined by the maximum possible order of the spectrum:

$$k_{\max} = c/\lambda. \qquad (8.24.2)$$

Substituting (8.24.2) into (8.24.1) yields

$$\delta = cN/\lambda. \qquad (8.24.3)$$

Since the product cN is the same for both gratings and the observed spectral lines are the same, the resolving power of the two gratings must also be the same. A small difference in resolving powers determined via (8.24.3) can be caused by the fact that the exact form of (8.24.2) must be

$$k_{\max} \leqslant c/\lambda, \qquad (8.24.4)$$

whence

$$\delta_{\max} \leqslant cN/\lambda. \qquad (8.24.5)$$

Since only the integral parts are taken in (8.24.4) and (8.24.5), the values of $\delta_{\max}$ of the two gratings may differ somewhat.

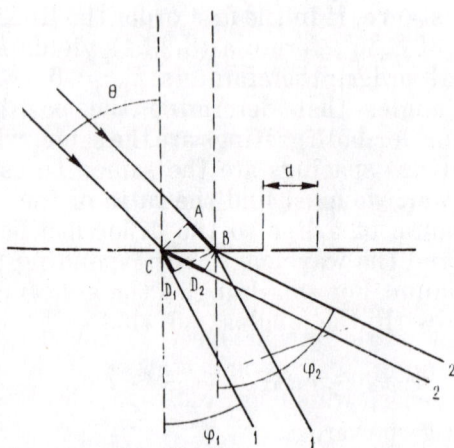

Fig. 8.25

8.25. The path difference between the rays from two neighboring slits is determined, as illustrated by the figure accompanying the answer, for direction *1* by the difference between the segments AB and CD_1 and for direction 2, by the difference between AB and CD_2. Accordingly, the path differences for directions 1 and 2 are

$$\delta_1 = d (\sin \theta - \sin \varphi_1) \quad \text{and} \quad \delta_2 = d (\sin \varphi_2 - \sin \theta),$$

or

$$\delta_1 = 2d \cos \left(\frac{\theta + \varphi_1}{2} \right) \sin \left(\frac{\theta - \varphi_1}{2} \right),$$

$$\delta_2 = 2d \cos \left(\frac{\theta + \varphi_2}{2} \right) \sin \left(\frac{\varphi_1 - \theta}{2} \right).$$

Thus, the diffraction maximum conditions can be written thus:

$$2d \cos \left(\frac{\theta + \varphi_1}{2} \right) \sin \left(\frac{\theta - \varphi_1}{2} \right) = k\lambda,$$

$$2d \cos \left(\frac{\varphi_2 + \theta}{2} \right) \sin \left(\frac{\varphi_2 - \theta}{2} \right) = k\lambda.$$

In the first approximation we can assume that $\theta + \varphi_1 \approx \varphi_2 + \theta \approx 2\theta$. Hence.

$$d \cos \theta \times (\theta - \varphi_1) \approx k\lambda, \quad d \cos \theta \times (\varphi_2 - \theta) \approx k\lambda.$$

$$(8.25.1)$$

This formulas have the same form as for the case of normal incidence of light on a grating with spacing $d \cos \theta$.

The maximum order of the spectrum in which the wavelength λ is observed is

$$k = d \cos \theta / \lambda,$$

while the longest wavelength ($k = 1$) is

$$\lambda = d \cos \theta.$$

The dispersive power can be conveniently expressed in terms of the angle with respect to the direction of the zeroth maximum, $\theta - \varphi_1$ and $\varphi_2 - \theta$. If by ψ we denote these differences, which are close in absolute value, we find that

$$\frac{d\psi}{d\lambda} \approx \frac{k}{d \cos \theta \cos \psi}.$$

At angles θ close to $90°$, the dispersive power of the grating may be considerably higher than for normal incidence of

299

light on the grating. However, the maximum dispersive power is $1/\lambda \cos \psi$, just as for normal incidence.

8.26. If we assume that the diffracted rays are reflected in the plane of the grating just like in a mirror (see the figure accompanying the answer), we arrive at a pattern similar to the one obtained in the answer to Problem 8.25.

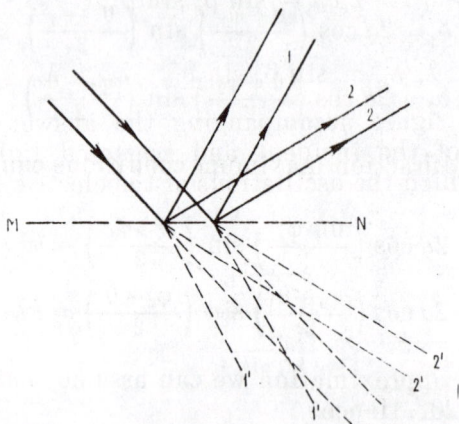

Fig. 8.26

Just like in the case of oblique incidence of the rays on the grating, the dispersive power increases with a coefficient of $(\cos \theta)^{-1}$.

8.27. According to Brewster's law, when light is reflected from a dielectric, complete polarization occurs when the tangent of the angle of incidence is equal to the refractive index of the medium reflecting the light. Since when light propagates in air and falls on a dielectric the refractive index is always greater than unity, we have $\tan \alpha > 1$, or $\alpha > 45°$.

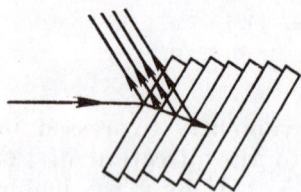

Fig. 8.28

8.28. Refracted light is polarized only partially. Light that is practically completely polarized can be obtained if one uses a stack (see the figure accompanying the answer) of parallel plates whose surfaces are oriented at the Brewster angle to the incident light. Light becomes partially polarized as it is refracted by the first plate, and as it travels from one plate to another, it becomes more and more polarized.

8.29. The ratio of the wavelengths is determined by the ratio of the speeds of propagation of the two waves:

$$\lambda_e/\lambda_0 \approx c_e/c_0.$$

At the same time,

$$c_e c_0 = \sin \beta_e \sin \beta_0.$$

Hence,

$$\lambda_e/\lambda_0 = \sin \beta_e/\sin \beta_0, \quad \lambda_e > \lambda_0.$$

8.30. The figure accompanying the answer shows the directions of the incident and scattered light and the planes in which the oscillations of the electric field vector

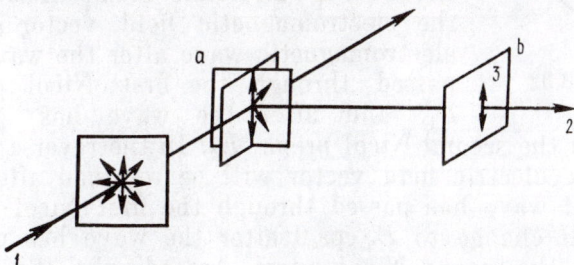

Fig. 8.30

lie. In the scattered light the oscillations must occur simultaneously in plane *a*, which is perpendicular to direction *1*, and in plane *b*, which is perpendicular to direction *2*. This, obviously, may happen only if the oscillations take place in the directions designated by arrow *3*. The blackening of the walls of the pipe, which was mentioned in the statement of the problem, is necessary so that no reflection can occur, since otherwise various directions of propagation of the light might become possible.

8.31. In the direction of the optic axis, the speed of propagation of the extraordinary and ordinary waves is the same and therefore the axis is perpendicular to the plane tangent to both wave surfaces at the point where the surfaces touch. In the first case (see Figure (a) accompanying the problem) the optic axis is parallel to the crystal boundary, while in the other (Figure (b)) it is perpendicular to the boundary. Since in all directions except the optic axis the speed of the extraordinary wave is higher than that of the ordinary, by the common nomenclature the crystal is negative.

8.32. After the light has passed through the first Nicol prism, its intensity becomes $I_1 = (1/2) I_0$ (it is assumed that the extraordinary wave loses no intensity when it is reflected and when it travels through a Nicol prism).

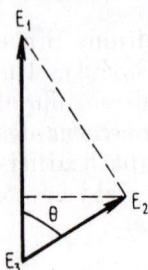

Fig. 8.32

According to Malus' law, after the light has passed through the second Nicol prism the intensity becomes

$$I_2 = I_1 \cos^2 \theta = (1/2) I_0 \cos^2 \theta.$$

The figure accompanying the answer shows the direction of oscillations of the electromagnetic field vector in the electromagnetic wave after the wave has passed through the first Nicol prism, E_1, and after the wave has passed through the second Nicol prism, E_2. In the reverse direction the electric field vector will be retained after the reflected wave has passed through the first Nicol prism but will change to $E_2 \cos \theta$ after the wave has passed through the second Nicol prism. Accordingly, the intensity after the light has passed through the two Nicol prisms in both directions will be

$$I_3 = I_2 \cos^2 \theta = I_1 \cos^4 \theta = (1/2) I_0 \cos^4 \theta.$$

8.33. The sense of rotation of the polarization plane depends on the direction of propagation of light in relation to the direction of the external magnetic field. For an overwhelming majority of substances ("positive" substances), the rotation is clockwise (looking in the direction of the ray of light) if the direction of propagation of light corresponds with that of the external magnetic field, and counterclockwise if the two directions are opposite. If the directions of the light ray and the external magnetic field coincided when the light passed from the source to the mirror and, therefore, the polarization plane rotated clockwise, after the light is reflected by the mirror the directions of the light ray and the external magnetic field are in opposition and the polarization plane rotates counterclockwise. If one views this process from the mirror, the rotation sense coincides with the clockwise rotation of the polarization plane when light passes in the primary direction. As a result, the two rotations are added and the angle doubles.

8.34. In the Kerr effect, the difference of the refractive indices of the extraordinary and ordinary waves obeys the law

$$n_e - n_o = kE^2, \qquad (8.34.1)$$

where k is a constant characterizing the medium. Since the electric field strength is squared in (8.34.1), the difference $n_e - n_o$ does not depend on the direction of the electric field. The optic axis in nitrobenzene coincides in direction with the electric field vector. The path difference between the ordinary and extraordinary rays,

$$\delta = l\,(n_e - n_o) = kE^2 l$$

(l is the length of the light path in the nitrobenzene), is also independent of the direction of the electric field vector, whereby the optical pattern caused by the emerging elliptical polarization will not change under reversal of direction of electric field.

8.35. According to classical theory, when a source of electromagnetic waves moves toward the observer, the ratio of the perceptible frequency to the frequency of the light emitted by a fixed source is

$$\frac{v_{cl}}{v_0} = \frac{1}{1-\beta} ,$$

with β the ratio of the speed of the source to the speed of light. According to the theory of relativity, this frequency ratio does not depend on whether the source or the observer is considered fixed and

$$\frac{v_{t.r}}{v_0} = \sqrt{\frac{1+\beta}{1-\beta}} .$$

The $v_{t.r}$-to-v_{cl} ratio is given by the formula

$$\frac{v_{t.r}}{v_{cl}} = \sqrt{1-\beta^2}.$$

Hence, the upper curve corresponds to classical-theory results, while the lower curve corresponds to the theory-of-relativity results. For $\beta \ll 1$ the difference between the two formulas is moderate (e.g. at $\beta = 0.1$ the difference amounts only to 0.5%).

8.36. The ratio of the ion velocity to the speed of light, $\beta = v/c$, in the case at hand is of the order of 10^{-4}. For such values of β the difference between the classical and relativistic formulas for the Doppler effect is negligible. If the source moves with a velocity v, the wavelength of the light measured by the receiver is

$$\lambda = \lambda_0 \pm vT = \lambda_0 \left(1 \pm v/c\right).$$

Here λ_0 is the wavelength of the light emitted by a fixed source, and the plus sign corresponds to the case where the source is moving away from the receiver, while the minus sign corresponds to the case where the source is moving toward the receiver. The difference in wavelengths measured from both sides of the tube with the plasma in which the ions move is

$$\Delta\lambda = 2\lambda_0 \left(v/c\right),$$

which yields the following formula for the velocity of the ions:

$$v = \frac{\Delta\lambda}{2\lambda_0} \, c.$$

Since the ions have different velocities, each observed spectral line is blurred, or broadened. The maximal intensity corresponds to the most probable velocity, while the extent to which the line is blurred characterizes the velocity distribution of the directional motion of the ions.

8.37. Since the velocities of atoms are much lower than the speed of light, we can employ the classical formulas for the Doppler effect. As shown in the answer to Problem 8.36, the difference in the wavelengths of the waves emitted by two identical sources that move with velocities of the same absolute value but pointing in opposite directions in relation to the receiver constitutes

$$\Delta\lambda = 2\lambda_0 \left(v/c\right),$$

where λ_0 is the wavelength of the wave emitted by a fixed source, and c is the speed of light. In a light-emitting gas, the atoms move with different velocities, in accordance with the Maxwellian distribution law. The higher the temperature, the more extended is the distribution in the direction of higher temperatures, therefore the higher the temperature, the broader the spectral line. Hence, curve 2 corresponds to a higher temperature.

8.38. In accord with the Doppler principle, the distribution in wavelengths of the intensity of the emission lines of excited ions reflects the velocity distribution of the ions (and hence the energy distribution of the ions, too). However, this distribution cannot be associated with the temperature of the gas. The fact is that the motion of ions in the discharge plasma (which is the source of radiation emitted in the tube) is highly anisotropic; this anisotropy is determined by the electric field strength in the tube. The electric field in the tube has a radial component directed from the axis to the wall. On the axis this component is zero; it increases as we approach the wall. This field imparts a directional velocity to the ions. Thus, the left half of the curve in Figure (b) (shorter wavelengths) corresponds to the ions moving away from the axis toward the spectrograph, while the right half corresponds to the ions moving away from the axis in the opposite direction.

8.39. According to Kirchhoff's law, the ratio of the total emissivity of a heat radiator to the absorption coefficient (immissivity) of that same radiator is the same for all objects, constitutes a universal function of the temperature, and is equal to the total emissivity of a black body:

$$e_\text{T}/a_\text{T} = E_\text{T}.$$

Hence, an object with a higher absorption coefficient has a higher emissivity and, therefore, it loses the energy acquired during heating at a higher rate. Curve 1 (see the figure accompanying the problem), therefore, represents the change of temperature in cooling for the object with the lower absorption coefficient or, in other words, curve 2 represents the cooling off of the object with the higher absorption coefficient.

8.40. The average kinetic energy of a molecule of the gas in translational motion is

$$w = \frac{3}{2}\, kT,$$

where k is the Boltzmann constant. If the concentration of the molecules in the gas is n, the volume density of the energy of the molecules is

$$u_\text{m} = \frac{3}{2}\, nkT.$$

The volume density of the energy of blackbody radiation, according to the Stefan-Boltzmann law, is

$$u_r = \frac{4\sigma}{c} T^4.$$

If we set u_m equal to u_r, we get

$$T = \left(\frac{3nkc}{8\sigma} \right)^{1/3} \tag{8.40.1}$$

We will illustrate the above result with two examples. First, suppose that the concentration of the molecules is the same as that at S.T.P. conditions ($T = 273$ K, $p = 101\ 325$ Pa). This concentration (the Loschmidt number) n is equal to 2.686×10^{25} m^{-3}. Substituting into (8.40.1) the values $k = 1.3807 \times 10^{-23}$ J/K, $c = 2.9979 \times 10^8$ m/s, and $\sigma = 5.670 \times 10^{-8}$ W$\cdot$m$^{-2}\cdot$K^{-4}, we find that

$$T = 9.03 \times 10^5 \text{ K}.$$

Under these assumptions, the gas pressure is

$$p = nkT = 3.35 \times 10^8 \text{ Pa} = 3300 \text{ atm}.$$

In the second example, we wish to find the concentration of the molecules of the gas if the temperature at which the energy density of the translational motion of the molecules is equal to the energy density of electromagnetic radiation is to be equal to 0 °C. Equation (8.40.1) yields

$$n = 7.42 \times 10^{14} \text{ m}^{-3}.$$

This concentration yields the following value for the pressure of the gas:

$$p = 2.8 \times 10^{-6} \text{ Pa}.$$

8.41. The emissive power over a definite wavelength interval is

$$\Delta E_T = \int_{\lambda_1}^{\lambda_2} E_{\lambda T}\, d\lambda.$$

Since the integral is the area under the curve limited by the ordinates corresponding to the lower and upper values, the emissive power per each interval is the same. The energy of the quanta corresponding to greater wavelengths is lower, whereby even for the same emissive

306

power there are more quanta of lower energy (i.e. referring to S_2).

8.42. Contrary to Wien's displacement law, the maximum in the blackbody radiation distribution corresponds, for a higher temperature, to a longer wavelength rather than to a shorter wavelength.

8.43. The relationship that exists between the radiation function and the volume radiation density is

$$E_{\nu T} = u_{\nu T} c/4.$$

The radiant emittance over the frequency range from ν_1 to ν_2 is determined by the integral

$$\Delta E_{1,2} = \int_{\nu_1}^{\nu_2} E_{\nu T}\, d\nu,$$

and, hence, the volume radiation density over the same range is

$$\Delta u_{1,2} = \frac{4}{c} \int_{\nu_1}^{\nu_2} E_{\nu T}\, d\nu.$$

8.44. The thermal radiation emitted by a body cannot exceed the blackbody radiation over all possible wavelength intervals. Contrary to this theoretical fact, the experimental curve contains a section that lies above the curve representing blackbody radiation.

8.45. According to Kirchhoff's law,

$$e_\lambda/a_\lambda = E_\lambda,$$

where E_λ and e_λ are the respective radiant emittance of

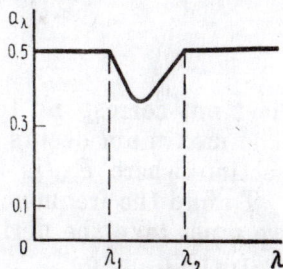

Fig. 8.45

a black body and a given object (which is not a black body), and a_λ is the absorption coefficient of the object. Therefore, the ratio of the ordinates of curve 2 to those of curve 1 yields the value of a_λ for each wavelength. On the segment from $\lambda = 0$ to λ_1 the value of a_λ remains constant and equal to 0.5. The same happens on the segment from λ_2 to $\lambda = \infty$. On the segment from λ_1 to λ_2

the value of a_λ passes through a minimum, as shown in the figure accompanying the answer.

8.46. Since

$$E_\nu = \frac{dE}{d\nu}, \quad E_\lambda = \frac{dE}{d\lambda}, \quad \text{and} \quad \frac{d\lambda}{d\nu} = -\frac{c}{\nu^2},$$

we have

$$E_\nu = \frac{dE}{d\lambda}\frac{d\lambda}{d\nu} = -\frac{c}{\nu^2}\frac{dE}{d\lambda} = -\frac{c}{\nu^2}E_\lambda.$$

To compare the maximal values of E_ν and E_λ, we take the derivative

$$\frac{dE_\nu}{d\nu} = \frac{2c}{\nu^3}E_\lambda - \frac{c}{\nu^2}\frac{dE_\lambda}{d\nu} = \frac{c}{\nu^3}\left(2E_\lambda + \frac{c}{\nu}\frac{dE_\lambda}{d\lambda}\right)$$

At the maximum of E_λ the second term is zero while the first is not. Thus, at the wavelength λ_m the frequency

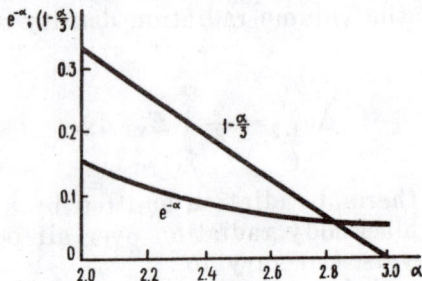

Fig. 8.46

does not correspond to the one at which E_ν is maximal. The maximum occurs at $dE_\lambda/d\lambda$ negative, that is, in the section where E_λ is falling off.

To find the frequency ν_m at which E_ν has its maximum, we must take the derivative of the Planck function with respect to ν, or

$$\frac{dE_\nu}{d\nu} = \frac{2\pi h}{c^2}\left\{\frac{3\nu^2\left[\exp\left(\frac{h\nu}{kT}\right)-1\right] - \frac{h\nu^3}{kT}\exp\left(\frac{h\nu}{kT}\right)}{\left[\exp\left(\frac{h\nu}{kT}\right)-1\right]^2}\right\}.$$

Nullifying this derivative, we arrive at a transcendental equation for $h\nu/kT$:

$$e^{h\nu/kT}\left(3 - \frac{h\nu}{kT}\right) = 3.$$

This equation can be solved graphically by constructing two functions,

$$y_1 = e^{-h\nu/kT} \quad \text{and} \quad y_2 = 1 - h\nu/3kT.$$

An approximate determination (via the intersection point of the two curves) yields a value of 2.82 for $h\nu_m/kT$. A more exact calculation yields

$$h\nu_m/kT = 2.8214,$$

or

$$h\nu_m = 3.896 \times 10^{-23} \ T. \qquad (8.46.1)$$

From (8.46.1) it follows that Wien's displacement law can be written in the form

$$\nu_m = 5.879 \times 10^{10} \ T.$$

The frequency ν_m corresponds to the wavelength (we denote it by $\lambda \, (\nu_m)$)

$$\lambda \, (\nu_m) = 5.10 \times 10^{-2} \ T^{-1}.$$

Thus,

$$\lambda \, (\nu_m)/\lambda_m = 1.760.$$

8.47. The volume density of the energy of blackbody radiation over the frequency range from ν to $\nu + d\nu$ is determined from the Planck formula

$$du = \frac{8\pi h\nu^3}{c^3} \frac{1}{\exp\left(\dfrac{h\nu}{kT}\right) - 1} \, d\nu.$$

The energy of each quantum in this range is $h\nu$. Thus, the distribution function for the number of quanta over the energy of one quantum has the form

$$\frac{dn}{dh\nu} = \frac{8\pi \, (h\nu)^2}{c^2 h^3} \frac{1}{\exp\left[h\nu/(kT)\right] - 1} \, . \qquad (8.47.1)$$

Introducing the dimensionless parameter $\alpha = h\nu/kT$, we can represent (8.47.1) in the form

$$n_\alpha = \frac{dn}{d\alpha} = \frac{8\pi k^3}{c^3 h^3} \, T^3 \, \frac{\alpha^2}{e^\alpha - 1} \, . \qquad (8.47.2)$$

The total "concentration" of the quanta can be obtained by integrating (8.47.2) with respect to α from zero to infinity, and the result is

$$n = \frac{8\pi k^3}{c^3 h^3} T^3 \int_0^\infty \frac{\alpha^2\, d\alpha}{e^\alpha - 1}. \qquad (8.47.3)$$

The integral in (8.47.3) can be reduced to tabulated functions (it can also be evaluated by expanding it in a power series). The value of the integral is 2.404, with the result that

$$n = \frac{8 \times 2.404 \pi k^3}{c^3 h^3} T^3 = 2.028 \times 10^7\, T^3.$$

In relative units of $(1/n)\, dn/d\alpha$, the energy distribution function for the quanta is presented in Figure (a).

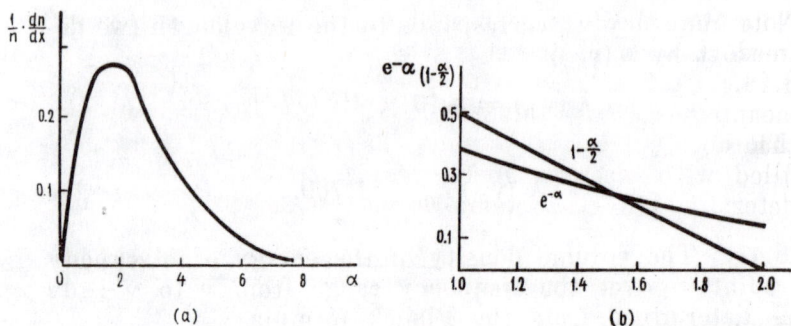

Fig. 8.47

Since the total energy density of blackbody radiation energy is

$$u = \frac{4\sigma}{c} T^4 = 7.57 \times 10^{-16} T^4$$

(σ is the constant in the Stefan-Boltzmann law), knowing the total number of quanta (see formula (8.47.3)) we can determine the average energy of a single quantum:

$$h\nu_{av} = \frac{7.57}{2.028} \times 10^{-3} T = 3.73 \times 10^{-23} T = 2.70 kT.$$

The distribution function given by (8.47.2) enables finding the energy of the "most probable" quantum, that is, the quantum whose energy corresponds to the maximum in the distribution function. To this end one must nullify

310

the derivative $dn_\alpha/d\alpha$. This leads to the transcendental equation

$$(2 - \alpha)\, e^\alpha = 2.$$

An approximate graphical solution (Figure (b)) yields $\alpha = 1.6$. A more exact value is $\alpha = 1.594$. Hence,

$$h\nu_p = 1.594kT.$$

In the answer to Problem 8.46 it was shown that the energy of the quantum corresponding to the maximum of the function $E_{\nu T}$ is $h\nu_m = 2.8214kT$. Wien's displacement law can then be used to determine the energy of the quantum corresponding to the maximum of the function:

$$h\nu\,(\lambda_m) = \frac{ch}{b}\, T = 6.855kT.$$

Note that the average kinetic energy per one degree of freedom of an ideal gas is $w = 0.5kT$.

8.48. At first glance it appears that the question is meaningless. Just think, how can one heat something that does not exist? Actually, however, space is always filled with electromagnetic radiation, whose energy is determined by the Stefan-Boltzmann law:

$$u = \frac{4\sigma}{c}\, T^4. \qquad (8.48.1)$$

If we imagine a region in space bounded by a shell that radiation cannot penetrate either from the outside or from within (and inside the shell a perfect vacuum is maintained), then the electromagnetic radiation inside the shell must be in thermodynamic equilibrium with the shell. To raise the temperature of the shell, we must supply an amount of heat determined not only by the heat capacity of the shell but also by the necessary increase in the density of energy of the electromagnetic radiation inside the shell. If we define the volume specific heat capacity as

$$c_{vol} = \frac{1}{V}\, \frac{dQ}{dT} = \frac{du}{dT}$$

and use formula (8.48.1) to find the derivative, we get

$$c_{vol} = \frac{16\sigma}{c}\, T^3.$$

8.49. If the intensity of the light is I, the number of photons of monochromatic light incident every second on a surface of unit area is

$$N = I/h\nu.$$

The momentum of each photon is $h\nu/c$. When hitting the surface, a photon transfers a momentum $h\nu/c$ to the surface if it is totally absorbed or a momentum $2h\nu/c$ if it is totally reflected. The pressure exerted on the surface is equal to the sum of all momenta transferred to the surface per unit time. In the case of absorption,

$$p = \frac{I}{h\nu}\,\frac{h\nu}{c} = \frac{I}{c},$$

while in the case of reflection,

$$p = 2\frac{I}{h\nu}\,\frac{h\nu}{c} = 2\,\frac{I}{c}.$$

If a fraction of the photons are absorbed and the rest are reflected, the latter process being characterized by a reflection coefficient R, then the pressure exerted by the light on the surface is

$$p = \frac{I}{c}\,(1 + R).$$

This formula coincides with (8.49.1), which was obtained on the basis of the electromagnetic theory of light.

8.50. Let us assume that such radiation has been obtained and is directed onto a mirror that is a paraboloid of revolution, with the

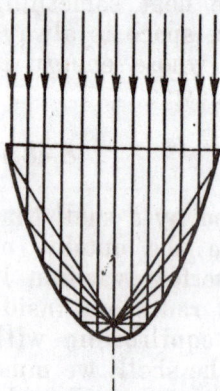

Fig. 8.50

rays of light being strictly parallel to the axis of the paraboloid (see the figure accompanying the answer). Since planes that are perpendicular to the rays are wave surfaces, all points in a single plane are in the same phase of oscillation (irrespective of the nature of the oscillation). All rays parallel to the axis converge (after being reflected) at a geometric point that is the focus of the paraboloid. The geometrical properties of a parabola imply that the sum of distances from any point in a plane that is perpendicular to the axis to the parabola and from the parabola to the focus is a constant. This means that the oscillations that arrive at the focus from all points in a wave surface are in phase. Hence, all radiation that travels to the paraboloid will be concentrated at a single point and the volume energy density of the radiation will become infinite at that point. This would make it possible to obtain (theoretically) infinite local temperatures at a finite temperature of the radiation source that provides the flow of plane waves.

The picture can be reversed, that is, we may ask ourselves: what requirements must a source meet for it to produce a stream of plane waves? Taking into account the reversibility of light rays, we conclude that such a source must be concentrated at a geometric point. At present quantum electronics can produce radiation with extremely low angular divergence, something on the order of 10^{-2} or even 10^{-3} of one second of the arc and, respectively, with colossal local power outputs. But even in this case the rays in such radiation cannot be considered strictly parallel.

8.51. The photon energy transferred to an electron in the metal is used to overcome the potential barrier at the boundary of the metal (the work function P) and part of it is lost inside the metal. In addition, one must bear in mind that not only the electrons that occupy levels lying near the Fermi level participate in the photoeffect. In addition to these, there are electrons that move somewhat slower and, hence, require for their liberation energies greater than the external work function. Therefore, Einstein's equation can be written in the form

$$hv = A + P + W,$$

where A is the term characterizing the energy losses inside the metal and the additional energy necessary for the electrons lying below the Fermi level to become liberated. The photoelectrons that escape from the surface of the metal have the maximal energy ($A = 0$); the initial energy of such electrons corresponds to the Fermi level:

$$W_\mathrm{m} = hv - P.$$

8.52. According to Einstein's equation,

$$hv = P + mv_\mathrm{m}^2/2,$$

313

where v_m is the maximal energy of the photoelectrons, and P is the work function of electrons ejected by the cathode. To stop the photoelectron current, we must apply a stopping potential no smaller than U_{stop}, which is determined from the equation

$$mv_m^2/2 = eU_{stop},$$

where e is the electron charge. Thus,

$$h\nu = P + eU_{stop}.$$

For a known value of e, the slope of the straight lines, $dU_{stop}/d\nu = h/e$, determines the Planck constant. The straight lines are different because they correspond to cathodes with different work functions. The work function can be determined either by the point of intersection of a straight line (for a particular cathode) with the horizontal axis,

$$P = h\nu_0$$

(with ν_0 the photoelectric threshold), or by the point of intersection of the straight line with the vertical axis,

$$P = -eU_{stop\,0}.$$

8.53. According to the hypothesis, the illuminated electrode emits photoelectrons whose maximal energy is

$$W_m = hc/\lambda - P,$$

which makes it possible to think of the system as an emf source, with the maximal value of the emf being

$$\mathcal{E} = W_m/e. \tag{8.53.1}$$

This source can generate a current in the circuit; the current is determined by the intensity of illumination of the electrode but cannot exceed a value of

$$I_m = \mathcal{E}/R.$$

At the same time, the current cannot exceed the value

$$I = Ne,$$

where N is the number of electrons ejected by the cathode per unit time due to illumination of the cathode with light. Since according to (8.53.1) the emf is constant and so is the value of R, the interelectrode gap may be considered as a resistance r_{vac} whose value is the smaller

314

the greater the intensity of the light. In darkness this resistance is infinite. Bearing all this in mind, we can write

$$I = \mathcal{E}/(R + r_{\text{vac}}).$$

8.54. The stopping potential difference, that is the voltage at which the photocurrent ceases, is the same for both cases. This potential difference determines the maximal photoelectron energy and equals the difference between the photon energy and the work function; hence, the emission frequency for the two sources is the same, and the sources differ only in the intensity of the radiation they emit.

8.55. According to Einstein's formula, the work function is equal to the difference between the photon energy and the maximal kinetic energy of the photoelectrons:

$$P = h\nu - mv_{\text{m}}^2/2.$$

The higher the maximal energy of the photoelectrons, which energy is equal to the maximal stopping potential, the lower the work function. In the case at hand, the cathode whose current-voltage characteristic is represented by curve 2 has a higher work function.

8.56. The point that an electron can reach thanks to their initial kinetic energy is determined only by the value of the stopping potential difference. Irrespective of the distance between the electrodes, the point is always at the middle of the interelectrode gap, and only such a distance can the fastest electrons leaving the cathode cover.

8.57. In Compton scattering, the photon wavelength changes by

$$\Delta\lambda = \frac{h}{m_e c}\,(1 - \cos\theta).$$

We see that in the case of angle θ_2 the wavelength increases by a larger quantity. Hence, $h\nu_2 < h\nu_1$. As a result of scattering, the photon transfers a fraction of its energy to the electron, and the energy that the electron receives is the greater, the smaller the energy of the photon after scattering, and hence the greater the value of θ is.

9. Atomic and Nuclear Physics

9.1. The protons move toward each other until their relative velocity becomes equal to zero. When the velocity is zero, the incident proton slows down and the immobile proton begins to accelerate, so that the distance between the two protons starts to increase. According to momentum conservation, when this happens, mv_0 becomes equal to $2mv$, where v is the velocity of both protons at the moment when the distance between the protons is minimal. At this moment both the velocities and, hence, the kinetic energies of the two protons are the same. The difference between the initial kinetic energy of the incident proton and the total kinetic energies of the two protons is equal to the energy associated with the interaction between the protons:

$$\frac{mv_0^2}{2} - 2\,\frac{m\,(v_0/2)^2}{2} = \frac{e^2}{4\pi\varepsilon_0 r}\,,$$

whence

$$r = \frac{e^2}{\pi\varepsilon_0 mv_0^2}.$$

9.2. Assuming that ionization occurs as a result of a completely inelastic collision, we can write

$$mv_0 = (m + m_H)\,u,$$

where m is the mass of the incident particle, m_H the mass of a hydrogen atom, v_0 the initial velocity of the incident particle, and u the final common velocity of the particle after collision. Prior to collision, the kinetic energy of the incident particle was

$$W_0 = mv_0^2/2.$$

The total kinetic energy after collision is

$$W = \frac{(m + m_H)\,u^2}{2} = \frac{m^2 v^2}{2\,(m + m_H)}\,.$$

The decrease in kinetic energy must be equal to the ionization energy:

$$W_0 - W = W_i = \frac{m_H}{m + m_H}\,W_0.$$

The greater the mass of the incident particle, the smaller the fraction of the initial kinetic energy that can be used for ionization. When an electron is used as the ionization agent, the initial kinetic energy of the electron is almost completely used for ionization. When an accelerated ion of hydrogen is used for ionization, the initial kinetic energy must double that of the electron, and when ionization is initiated by a helium atom, the energy must be five times that of the electron. This estimate explains why in a gas-discharge plasma, ionization is initiated almost exclusively by electrons, while ionization by the proper ions plays practically no role.

9.3. The kinetic energy of the electron in a hydrogen-like atom is

$$W_{kin} = \frac{me^4 Z^2}{8\varepsilon_0^2 n^2 h^2},$$

while the potential energy is

$$W_{pot} = -\frac{me^4 Z^2}{4\varepsilon_0^2 n^2 h^2}$$

As n grows (i.e. as the electron moves to higher levels), W_{kin} decreases in inverse proportion to n^2, while W_{pot} grows, tending to the maximal value of $W_{pot} = 0$ as $n \to \infty$. The total energy,

$$W = -\frac{me^4 Z^2}{8\varepsilon_0^2 n^2 h^2},$$

also tends to zero as $n \to \infty$. The minimal value of the total energy is

$$W_{min} = -\frac{me^4 Z^2}{8\varepsilon_0^2 h^2}.$$

Obviously, to detach the electron from the atom, the following work must be performed:

$$A = W_{max} - W_{min} = 0 - \left(-\frac{me^4 Z^2}{8\varepsilon_0^2 h^2} \right) = \frac{me^4 Z^2}{8\varepsilon_0^2 h^2}.$$

The ratio of this quantity to the elementary charge e is known as the ionization potential. This is the minimal potential difference that a particle of infinitely small mass and carrying the elementary charge (practically an electron) must pass for the given atom to become ionized.

9.4. The wave number of the emission lines of a hydrogen-like atom (when an electron "travels" from one quantum level to another) is given by the formula

$$\tilde{\nu} = RZ^2 \left(\frac{1}{k^2} - \frac{1}{n^2} \right),$$

where R is the Rydberg constant. For a nucleus of infinite mass,

$$R_\infty = \frac{me^4}{8\varepsilon_0^2 h^2 c}.$$

For a nucleus that has a finite mass M we must substitute the reduced mass

$$\mu = \frac{m}{1 + m/M}$$

for the electron mass m. Assuming that the electron energy is zero at infinity, we arrive at the following formula for the energy level with the principal quantum number n:

$$W_n = - R_\infty \frac{ch}{n^2} \frac{1}{1 + m/M}.$$

This formula shows that the greater the mass of the nucleus, the deeper are the levels of the nucleus and the greater the separation of the levels and the higher the frequency of the spectral line reflecting the transition between levels with the same initial quantum numbers and the same final quantum numbers. Of course, since $m \ll M$, the difference between the corresponding values is small, but for hydrogen and deuterium it is sufficiently high. The aforesaid implies that system 1 belongs to deuterium and system 2, to hydrogen.

9.5. An ionized helium atom belongs to the class of atoms known as hydrogen-like, for which the following general series formula is valid:

$$\tilde{\nu} = R_M Z^2 \left(\frac{1}{k^2} - \frac{1}{n^2} \right),$$

where Z is the proton number. The Rydberg constant for an atom whose mass is M is

$$R_M = R_\infty \frac{1}{1 + m/M}. \qquad (9.5.1)$$

If we ignore the difference between the Rydberg constants for hydrogen and a helium ion, then it can be assumed that the lines of the first coincide with those of the second. This occurs if

$$4 \left(\frac{1}{k_{He}^2} - \frac{1}{n_{He}^2} \right) = \frac{1}{k_H^2} - \frac{1}{n_H^2}.$$

In the Balmer series, $k = 2$. We set $n_H = n_{He} = \infty$. Then

$$k_{He} = 4 \quad \text{and} \quad n_{He} = 6, 8, 10, 12, \ldots.$$

In the spectrum of a helium ion, between these lines are the lines for which $n_{He} = 5, 7, 9, 11, \ldots$ These lines are also shown in the figure accompanying the problem. We note, in connection with formula (9.5.1), that since $R_{He} > R_H$, the lines of a helium atom correspond to slightly higher frequencies than the corresponding lines in the Balmer series.

9.6. For a doubly ionized lithium atom, $Z = 3$. For this reason the spectral lines of the lithium ion are described by the general series formula

$$\tilde{\nu} = 9R \left(\frac{1}{k_{Li}^2} - \frac{1}{n_{Li}^2} \right).$$

For the Balmer series we have $k_H = 2$, whereby only the lines of lithium that obey the relationship $9/k_{Li}^2 = 1/4$ can be found in the visible spectrum. Hence

$$k_{Li} = 6.$$

The last line in the Balmer series corresponds to a value of the principal quantum number n_H being equal to 6. The corresponding line for lithium exists at $9/n_{Li}^2 = 1/6^2$, that is, at

$$n_{Li} = 18.$$

Thus, in the spectral region of the first four lines of the Balmer series the overall number of lines is 12 ($n_{Li} = 7, 8, 9, 10, 11, 12, 13, 14, 15, 16, 17, 18$). The lines with $n_{Li} = 9, 12, 15, 18$ lie close to the lines in the Balmer series with $n_H = 3, 4, 5, 6$. Since there is a small difference in the values of the Rydberg constant, these lines do not coincide exactly. The difference is somewhat greater than in the case of the Pickering series.

9.7. The electric field in which the electron is moving is

$$E = \frac{e}{4\pi\varepsilon_0 r^2}, \qquad (9.7.1)$$

where r is the radius of the electron orbit according to the "classical" Bohr theory. In the ground state of the hydrogen atom, the radius of the orbit is $r_1 = 5.29 \times 10^{-11}$ m. Formula (9.7.1) then yields the following value for the electric field strength:

$$E = 5.15 \times 10^{11} \text{ V/m},$$

which exceeds all practically attainable field strengths by several orders of magnitude. However, if an electron is moving along a circular orbit which corresponds to a value of the principal quantum number that differs from unity, the radius of such an orbit is

$$r = r_1 n^2,$$

and the electric field strength proves to be inversely proportional to n^2. If, say, $n = 10$, the electric field lies within the limits of practically attainable fields. Indeed, the ionization of highly excited states of the hydrogen atom by an electric field was actually observed in experiments.

9.8. Optical transitions between the ground state of helium and the 2^1S and 2^3S states are forbidden by selection rules. Although the selection rules that forbid such transitions are not absolute, they nevertheless permit defining the 2^1S and 2^3S states as metastable with lifetimes of the order of 10^{-3} s, which is an extremely large time interval on the scale of atomic processes. Excitation to such levels is possible in a discharge almost exclusively due to electron impact. What is needed for continuous generation of radiation is inverted population of levels. This becomes possible if the lifetime on the higher level exceeds considerably the lifetime on the lower level, with the result that the lower level has time to "get rid" of the electrons before new electrons arrive. Indeed, the lifetime of the $2S$ and $3S$ atomic states is of the order of 10^{-6} s, while the lifetime of state $2P$ is of the order of 10^{-8} s. In the first of the two transitions $3S \rightarrow 2P$ and $2S \rightarrow 2P$ the energy changes by a larger amount; hence a quantum of a higher frequency corresponds to this transition, and this frequency lies in the visible spectrum ($\lambda = 632.8$ nm), while the second transition corresponds to a quantum with a lower frequency, $\lambda = 1153$ nm, which lies in the IR region.

9.9. Since the length of all the vectors is the same, the absolute values of the angular momenta in all the states are the same, too. If the orbital quantum number is l, the magnetic quantum number m may assume $2l + 1$ different values. The figure accompanying the problem shows five different states. Hence, $l = 2$. The value of l cannot exceed $n - 1$, whereby the minimal value of the principal quantum number is 3. The values -2, -1, 0, $+1$, $+2$ of the magnetic quantum number correspond to different orientations of the angular momentum vector.

9.10. In a uniform magnetic field, a magnetic dipole, which is an object possessing a magnetic moment, experiences only a torque. For a force to act on a magnetic dipole, the field must be nonuniform. For an atomic magnetic moment this force is defined by the expression

$$F = \mu \frac{dB}{dz}, \qquad (9.10.1)$$

where μ is the magnetic moment of the atom. In formula (9.10.1) we assume that the vector of magnetic induction of the magnetic field generated by the atom is oriented along the lines of force of the external magnetic field and its direction coincides with that of the induction **B** of the external magnetic field or is opposite. In the first case the atom is pulled into the region where the field is stronger, while in the second case it is pushed out of that region. In the Stern-Gerlach experiment, the beam of silver atoms is sent through the (nonuniform) magnetic field and splits into two beams in accordance with two possible directions of the magnetic moment of a silver atom. If there was no spatial quantization, the silver atom would be oriented at random and the beam would spread in all directions. The silver atoms in the beam are in the ground state, whereby the difference in orientation is due to the different directions of the magnetic moment of outer electrons in silver atoms.

9.11. The minimal wavelength in the X-ray spectrum is determined by the maximal energy which a bombarding electron may transfer to the anode. This energy is eU and, hence,

$$\lambda_{min} = \frac{ch}{eU}.$$

If the voltage is decreased three-fold, the minimal wavelength increases three-fold, too. As the figure accompanying the problem shows, as a result of such an increase in the wavelength the short-wave peak, which is one of the characteristics of the material of the anode, disappears. Separate characteristic peaks may disappear even when the wavelength corresponding to these peaks is longer than λ_{min} if to excite the quantum level from which the transition that generates the radiation with the wavelength of a particular peak begins an energy higher than eU_0 is required.

9.12. In infinitely deep potential well, the wave function at the boundary of the well ($x = 0$ and $x = l$) is zero. Since the figure accompanying the problem clearly shows that the wave function does not vanish at the boundary, we conclude that the well is of finite depth.

9.13. In a potential well of infinite depth the wave function at the "walls" of the well must vanish. This means that only states labeled by even numbers, e.g. 2, 4, 6, etc., may remain. The distance between the nodes of a standing wave function is equal to one-half of the de Broglie wavelength:

$$\frac{\lambda}{2} = \frac{h}{2mv}.$$

The maximal value of λ is a, which means that the electron velocity has a minimal value $v = h/2ma$, and hence the minimal value of the electron energy is $W_{min} = h^2/8ma^2$. If the width of the well decreases two-fold, the minimal kinetic energy of the electron in the well increases four-fold.

9.14. If the initial kinetic energy of the electron in the motion from left to right is E, to the right of the barrier it will be $E - P$. In the first case the de Broglie wavelength is

$$\lambda_1 = h/\sqrt{2mE},$$

while in the second it is

$$\lambda_2 = h/\sqrt{2m\,(E-P)}.$$

The wavelength ratio is in inverse proportion to the refractive index ratio:

$$\frac{n_2}{n_1} = \frac{\lambda_1}{\lambda_2} = \sqrt{\frac{E-P}{E}}.$$

The right region can be considered as being less optically dense, whereby when the electron is moving from left to right the phase is retained, while when the electron is moving from right to left, the phase changes to its opposite.

9.15. From the viewpoint of classical mechanics, for $E < P$ this probability is zero in both cases, while for $E > P$ it is equal to unity ("step" 1 in Figures (a) and (b) accompanying the answer). From the viewpoint of quantum mechanics, however, in the first case for $E < P$

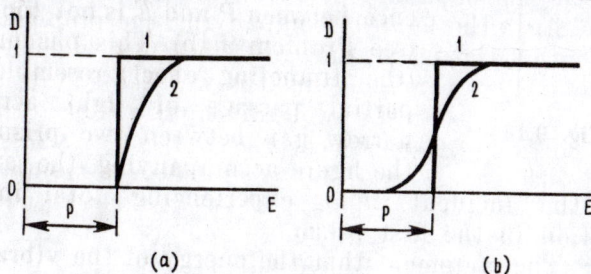

Fig. 9.15

the probability is also zero, whereas for $E > P$ the probability is lower than unity (curve 2 in Figure (a) accompanying the answer), since there is a nonzero probability of the electrons being reflected from the step, in other words, a fraction of the electrons moving from left to right begins to move in the opposite direction. Partial reflection takes place even when the potential energy to the left of x_0 is greater than the potential energy to the right of x_0 rather than lower. For the potential barrier depicted in Figure (b) accompanying the problem there is a nonzero probability of the electrons tunneling through the barrier even when $E < P$, but this probability does not become equal to unity even when $E > P$ (curve 2 in Figure (b) accompanying the answer). The passage of electrons through the potential barrier when $E < P$ under the conditions that the barrier has a finite width and that the potential energy to the right of the barrier is equal to or less than to the left of the barrier became known as the tunneling effect. This effect is encountered in many atomic and nuclear processes and in the field emission of electrons by metals and semiconductors. The probability of electrons passing through the barrier for $E < P$ is the higher the lower and narrower the barrier.

9.16. In region *II*, the wave function does not obey the sinusoidal law; it falls off exponentially. This happens within the framework of classical mechanics when a negative kinetic energy is assigned to the electron, or $E < P$. The passage of the electron into region *III*, which is forbidden from the classical standpoint, can be observed in experiments if the width of region *II* is sufficiently small (of the order of the electron wavelength in region *I*) and if the difference between P and E is not too great (see Problem 9.15). This phenomenon (the tunneling effect) resembles the partial passage of light across a narrow gap between two prisms (see the figure accompanying the answer) with the incident light experiencing total internal reflection in the first prism.

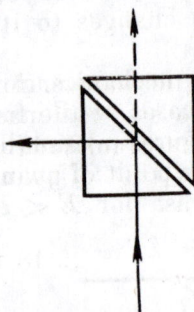

Fig. 9.16

9.17. The statement that the energy of the vibrational motion of atoms or molecules in a crystal lattice is nil at absolute zero contradicts one of the main principles of quantum mechanics, the uncertainty principle. If the kinetic energy is zero, so is the momentum. But if an atom or a molecule is at rest, its position is fixed. In other words, each coordinate and the projection of momentum on the respective coordinate axis are known with absolute accuracy. Meanwhile the wave properties of particles permit determining the collection of a coordinate and the respective projection of momentum within the intervals Δp_x and Δx, where in accordance with the uncertainty principle

$$\Delta p_x \Delta x \geqslant h/2\pi.$$

For this reason the energy of the atoms or molecules of a crystal is not nil at absolute zero. The motion of these objects is vibrational (zero-point vibrations), and the energy associated with this motion is the zero-point energy

$$E_0 = \frac{1}{2} h\nu = \frac{h}{4\pi} \omega,$$

where ω is the natural cyclic frequency of the vibration of a particle in the lattice. The existence of zero-point vibrations has been proved in experiments. They manifest themselves in light scattering in crystals at temperatures close to absolute zero.

9.18. The diffraction of electrons by a crystal obeys the same Bragg law as X-ray diffraction does:

$$2d \sin \theta = k\lambda.$$

In this formula $\lambda = h/mv$ is the de Broglie wavelength. Substituting the necessary constants (the electron mass and charge and the Planck constant) and transforming the units of measurement, we arrive at the following formula*:

$$\lambda = \sqrt{1.5/U} \text{ nm.} \qquad (9.18.1)$$

According to this formula, diffraction maxima are observed for the following wavelengths: λ_0 $(k = 1)$, $(1/2)\,\lambda_0$ $(k = 2)$, $(1/3)\,\lambda_0$ $(k = 3)$, etc., with the voltages that determine the electron energy being U_0, $U_0\sqrt{2}$, $U_0\sqrt{3}$, etc. If on the horizontal axis we lay off the square roots of the values of the accelerating voltage, as is done in Figure (b) accompanying the problem, the current maxima must be spaced by equal distances. In experiments, however, this condition is not met exactly, and the smaller the voltage the greater the deviation from this pattern. The reason for this is that formula (9.18.1) contains the energy (in electron volts) of an electron inside the metal, and this quantity is the sum of the energy acquired by the electron in passing the potential difference and the difference in potential energies of the electron inside and outside the metal. Therefore, along the horizontal axis in Figure (b) accompanying the problem we must lay off $\sqrt{U + \Phi}$ rather than $\sqrt{U}$, where Φ is the internal potential in the metal. The quantity measured in experiments is, of course, U. Electron diffraction patterns obtained as a result of electron scattering on a metal lattice make it possible to obtain Φ.

* Here U is the potential difference through which the electron travels and, hence, the electron energy expressed in electron volts.

9.19. The stability of a nucleus is ensured by the fact that the Coulomb repulsive force experienced by each proton in the nucleus is equal to the force of nuclear attraction (the nuclear force). The Coulomb force falls off with distance relatively slowly (in inverse proportion to the square of the distance), while the nuclear force falls off very rapidly. For this reason the protons are held in the nucleus only by the closest neutrons, while experiencing the repulsive action of all the protons in the nucleus, even those farthest from a given proton. Thus, as the general number of nucleons grows, more and more neutrons are required so as to compensate for the growing action of the Coulomb repulsive forces.

9.20. According to the Pauli exclusion principle, a single quantum level can carry no more than two identical particles with half-integral spin. The directions of the spins must be opposite. In a nucleus such particles are the nucleons, protons and neutrons. Since these are distinct particles, there can be not more than four nucleons on the lowest level—two neutrons and two protons.

9.21. If N_0 is the number of radioactive atoms in the radioactive sample at the beginning of counting and λ

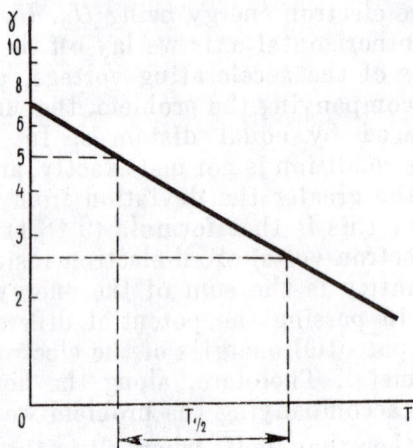

Fig. 9.21

is the decay constant, then at time t after the beginning of counting the number of atoms will be

$$N = N_0 e^{-\lambda t}. \qquad (9.21.1)$$

The rate with which this number changes is

$$\frac{dN}{dt} = -\lambda N_0 e^{-\lambda t} = -\lambda N.$$

A counter registers only the radioactive particles that

fly in its direction. The fraction of such particles in the overall number of radioactive particles emitted by the sample depends on the size and position of the counter and can be characterized by a factor a (with $a < 1$). Thus, the counting rate can be expressed in the form

$$\gamma = a \left| \frac{dN}{dt} \right| = aN_0 \lambda e^{-\lambda t}.$$

Taking logs, we get

$$\log \gamma = \log (aN_0 \lambda) - \lambda t.$$

To determine the half-life of the radioactive element, there is no need to measure the slope and find the λ vs. t dependence and, using the well-known formula, to calculate $T_{1/2}$. Suffice it to lay off in any place on the vertical axis a segment equal to the logarithm of two (irrespective of what logarithms are laid off on the vertical axis, base-10 or base-e) and draw through the end points of this segments straight lines parallel to the horizontal axis. The points at which these straight lines intersect the experimental straight line that represents the variation in the rate of counting determine the boundaries of the time interval in the course of which the counting rate decreases by a factor of 2. Since the experimental law representing the decrease in the counting rate with the passage of time coincides with the law representing the decrease in the number of radioactive atoms (9.21.1), this time interval is the sought half-life.

9.22. A shift to the right by one place in the Periodic Table occurs as a result of a beta decay act. The mass number does not change in this act while proton number increases by unity. Hence,

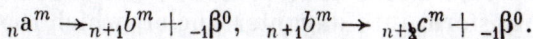

$$_n a^m \rightarrow {}_{n+1} b^m + {}_{-1}\beta^0, \quad _{n+1} b^m \rightarrow {}_{n+2} c^m + {}_{-1}\beta^0.$$

A shift to the left by two places occurs in alpha decay. The mass number decreases by four, while the proton number decreases by two:

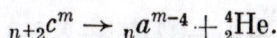

$$_{n+2} c^m \rightarrow {}_n a^{m-4} + {}_2^4 \text{He}.$$

The mass number of the resulting isotope of atom a differs from the initial number by four units.

Examples of such radioactive transformations are the

chains of transformations in the $^{238}_{92}$U and $^{232}_{90}$Th families:

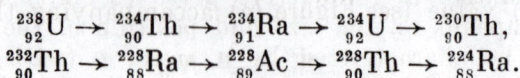

$$^{238}_{92}U \rightarrow {}^{234}_{90}Th \rightarrow {}^{234}_{91}Ra \rightarrow {}^{234}_{92}U \rightarrow {}^{230}_{90}Th,$$
$$^{232}_{90}Th \rightarrow {}^{228}_{88}Ra \rightarrow {}^{228}_{89}Ac \rightarrow {}^{228}_{90}Th \rightarrow {}^{224}_{88}Ra.$$

9.23. In the course of the time interval dt the number of nuclei of the new element (the "daughter" nuclei) changes thanks to the emergence of new nuclei as a result of the decay of initial (or "parent") nuclei and the departure of new nuclei as a result of their decay:

$$dN_2 = N_1\lambda_1 dt - N_2\lambda_2 dt.$$

Here N_1 is the number of parent nuclei and N_2 is the number of the daughter nuclei at the given moment. According to the law of radioactive decay,

$$N_1 = N_0 e^{-\lambda_1 t}.$$

Thus,

$$dN_2 = \lambda_1 N_0 e^{-\lambda_1 t}\, dt - \lambda_2 N_2\, dt,$$

or

$$\frac{dN_2}{dt} + \lambda_2 N_2 = \lambda_1 N_0 e^{-\lambda_1 t}. \qquad (9.23.1)$$

We start by considering the limiting cases.

(1) $\lambda_1 \gg \lambda_2$. If we rewrite (9.23.1) in the form

$$\frac{d(N_2/N_0)}{dt} + \lambda_2 \frac{N_2}{N_0} = \lambda_1 e^{-\lambda_1 t}$$

and assume that after a small time interval we can set $e^{-\lambda_1 t} = 0$, we obtain

$$\frac{N_2}{N_0} = \frac{N_{20}}{N_0} e^{-\lambda_2 t}.$$

With $\lambda_1 \gg \lambda_2$ we can assume that $N_{20} = N_0$, so that

$$N_2 = N_0 e^{-\lambda_2 t}.$$

Physically this means that parent nuclei practically instantly transform into daughter nuclei, which then decay according to the law of radioactive decay with a certain decay constant.

(2) $\lambda_1 \ll \lambda_2$. In this case the number of parent nuclei can be assumed to remain constant over a sizable time interval and is equal to N_0. This transforms (9.23.1) into

$$\frac{dN_2}{dt} = -(\lambda_2 N_2 - \lambda_1 N_0),$$

which after integration yields

$$N_2 = \frac{\lambda_1}{\lambda_2} N_0 (1 - e^{-\lambda_2 t}).$$

The number of daughter nuclei tends to a constant (saturation) value (see Figure (a) accompanying the answer):

$$N_2 = \frac{\lambda_1}{\lambda_2} N_0. \qquad (9.23.2)$$

Of course, over a long time interval this number will decrease in accord with the decrease of the number of parent nuclei, whereby a more exact form of (9.23.2) is

$$N_2 = \frac{\lambda_1}{\lambda_2} N_0 e^{-\lambda_1 t}.$$

An example of the case with $\lambda_1 \ll \lambda_2$ is the radioactive decay of radium $^{226}_{88}$Ra with a decay constant equal to 1.354×10^{-11} s^{-1} (a half-life of 1622 years). Its product

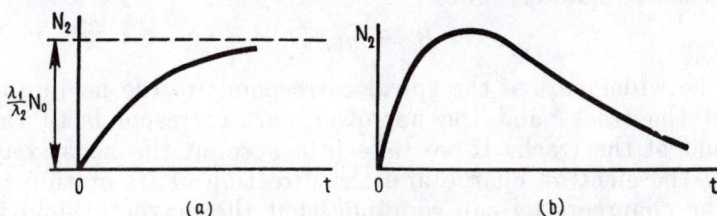

Fig. 9.23

is radon $^{226}_{86}$Rn with a decay constant equal to 2.097×10^{-6} s^{-1} (a half-life of 3.825 days). If radium is placed inside a closed vessel, already after one month the amount of radon in the vessel will be only 0.4% less than the equilibrium amount, while the equilibrium amount, as shown by (9.23.2), constitutes only 6.46 parts to a million of the initial number of radium atoms.

To find the overall dependence of N_2 on t, we must integrate Eq. (9.23.1). The solution has the form*

$$N_2 = \frac{\lambda_1 N_0}{\lambda_2 - \lambda_1} (e^{-\lambda_1 t} - e^{-\lambda_2 t}).$$

This expression has a maximum at a value of t equal to t_m, which can be found if we nullify the derivative dN_2/dt:

$$t_m = \frac{\ln \lambda_2 - \ln \lambda_1}{\lambda_2 - \lambda_1}.$$

The N_2 vs. t curve is depicted in Figure (b) accompanying the answer.

* To integrate Eq. (9.23.1), we introduce a new variable, $z = N_2 e^{\lambda_2 t}$. This yields

$$\frac{dz}{dt} = \left(\frac{dN_2}{dt} + \lambda_2 N_2 \right) e^{\lambda_2 t}, \quad e^{-\lambda_2 t} \frac{dz}{dt} = \lambda_1 N_0 e^{-\lambda_1 t}$$

$$dz = \lambda_1 N_0 e^{(\lambda_2 - \lambda_1)t} \, dt, \quad z = \frac{\lambda_1 N_0}{\lambda_2 - \lambda_1} (e^{-(\lambda_2 - \lambda_1)t} - 1),$$

$$N_2 = \frac{\lambda_1 N_0}{\lambda_2 - \lambda_1} (e^{-\lambda_1 t} - e^{-\lambda_2 t}).$$

9.24. As the electron moves in the Wilson chamber, it gradually loses its energy to ion formation, and it is on these ions that drops of mist form, which make visible the track of the electron. This loss of energy results in a loss of speed, which means that the radius of curvature of the electron trajectory in the external magnetic field becomes smaller, since

$$R = mv/eB.$$

The wider part of the spiral corresponds to the beginning of the track, and the narrower part corresponds to the end of the track. If we take into account the negativity of the electron charge and the direction of its motion in the chamber, we can conclude that the magnetic field is directed toward the reader.

9.25. According to Pauli's hypothesis, which was verified in experiments, simultaneously with the escape of an electron the nucleus emits a neutrino (more precisely, an antineutrino), which is the particle that carries off a fraction of the energy released in beta decay and which has a momentum whose vector sum with the nucleus momentum and the electron momentum is zero:

$$_z X^A \rightarrow _{z+1} Y^A + _{-1}\beta^0 + \tilde{\nu}.$$

9.26. The proton and neutron masses can be considered practically equal. When the proton and the neutron collide, the scattering angle after collision will be $90°$, whereby after collision the direction of the neutron velocity will also make an angle of $45°$ with the initial direction of the proton velocity. Thus, after collision the proton and neutron energies are practically the same.

9.27. A change in the direction of motion (following a collision act) by an angle greater than $90°$ is possible if the mass of the incident particle is smaller than that of the particle that initially was at rest (in the laboratory system). The mass of the atom and molecule of hydrogen is smaller than the mass of an alpha particle, while the mass of helium is equal to the mass of an alpha particle. The gas closest to helium in the Periodic Table that has a mass greater than that of the alpha particle is nitrogen.

9.28. The relative velocity, according to the relativistic formula for velocity addition, is

$$v_{rel} = \frac{v_1 + v_2}{1 + v_1 v_2 / c^2} \cdot$$

The velocity of the electron flying away from the accelerator with respect to the accelerator is

$$V_a = \frac{2v}{1 + v^2/c^2} \, ,$$

while the velocity of the electron flying toward the accelerator is

$$V_b = 0.$$

The relative velocities of the electrons with respect to each other are:

$$V_{ab} = \frac{2v}{1 + v^2/c^2}$$

for the electron moving away from the accelerator, and

$$V_{ba} = -\frac{2v}{1 + v^2/c^2}$$

for the electron moving toward the accelerator, that is, they are equal in absolute value.

For the sake of an example we assume that $v = 0.9c$ In this case the velocity of the electron flying away from the accelerator and the relative velocities of the electrons are related through the following formula:

$$V_a = V_{ab} = -V_{ba} = \frac{1.8}{1 + 0.81} c = 0.9945c.$$

9.29. The statement carries no physical meaning whatsoever. First, there is not a single physical quantity that can transform into another physical quantity (time cannot transform into area, field strength into length, and so on). Second, for processes in relation to which this statement is usually made, the common conservation laws, the energy conservation law and the mass conservation law, are valid, that is, if isolated systems are considered. In the case at hand, these balance equations are as follows:

$$E^* = E + h\nu$$

for energy (here E^* is the energy of the excited atom, E is the energy of the atom in the ground state, and $h\nu$ is the photon energy), and

$$m^* = m + \mu_0$$

for mass (here m^* is the mass of the excited atom, m is the mass of the atom in the ground state, and $\mu_0 = h\nu/c^2$ is the photon "mass").

The first balance equation expresses the law of energy conservation and the second, the law of mass conservation (for the same process).

9.30. The ratio of the mass of a moving particle to the rest mass of that particle is

$$\frac{m}{m_0} = \frac{1}{\sqrt{1 - v^2/c^2}}.$$

The kinetic energy acquired by a particle in an accelerator is determined by the following difference:

$$W_{kin} = mc^2 - m_0 c^2 = m_0 c^2 \left(\frac{1}{\sqrt{1 - v^2/c^2}} - 1 \right),$$

whence

$$\frac{m}{m_0} = \frac{W_{kin}}{m_0 c^2} + 1.$$

For a fixed value of W_{kin}, the ratio m/m_0 is the smaller the greater m_0 is and, hence, curve 2 corresponds to the particle with the greater rest mass.

9.31. If the kinetic energy of the particle is W_{kin}, its velocity can be found from the equation

$$W_{kin} = m_0 c^2 \left(\frac{1}{\sqrt{1 - v^2/c^2}} - 1 \right),$$

with the result that

$$\frac{v^2}{c^2} = 1 - \left(\frac{m_0 c^2}{W_{kin} + m_0 c^2} \right)^2.$$

If $W_{kin} \ll m_0 c^2$, we arrive at an expression for the velocity that is identical to the one following from classical mechanics and electrodynamics:

$$v = \sqrt{2eU_0 N / m_0}.$$

The voltage across the cylinders changes its sign in the course of a half-period $T/2 = 1/2v$, whereby the length of the cylinders must increase according to the law

$$l_N = \frac{1}{2v} \sqrt{\frac{2eU_0}{m_0}} \, N^{1/2}.$$

However, as W_{kin} grows, the velocity grows slower and slower. For instance, for $v = 0.87c$, $v = 0.89c$, and $v = 0.90c$ we have, respectively, $W_{kin} = m_0 c^2$, $2m_0 c^2$, and $3m_0 c^2$. For sufficiently high energies the velocity of the particle approaches that of light and the length of the cylinders does not change any more: $l = c/v$.

9.32. The operation of a cyclotron is based on the fact that the time a charged particle takes to perform a full circle in a magnetic field does not depend on the particle's velocity. The time it takes the particle to complete one-half of a full circle, that is, the time in the course of which the electric field between the Dees reverses its direction, is $\pi m v / Be$. As the particle is accelerated, its mass grows according to the formula

$$m = \frac{m_0}{\sqrt{1 - v^2/c^2}}.$$

The particle moving inside a Dee will gradually begin to get out of step with oscillatory electric field between the Dees.

The electron mass is doubled already at an energy equal to 0.51 MeV, whereby the discrepancy between the time it takes the electron to make a half-circle and the period of reversal of direction of the field between the Dees becomes noticeable already at accelerating voltages of the tens of kiloelectronvolts. This, naturally, limits the possibility of accelerating to high energies electrons in cyclotrons.

For ions, whose rest mass is greater than the electron rest mass by a factor of 10^3, 10^4 or even 10^5, the effect of increase of mass with velocity manifests itself at much higher energies. But here, too, there is a limit of acceleration of such particles in a cyclotron. To overcome this difficulty, other types of accelerators have been designed, in which the frequency of the electric field or the magnetic field is varied in the proper manner (separately or together).

9.33. The energy of the quantum that flies upward decreases while that of the quantum flying downward increases, as a result of which the frequency of the first gets lower and that of the second, increases. The difference proves to be so small, that could be detected only after a discovery made by Mössbauer, whose name was later given to this effect. An experiment in "weighing" the photon was conducted later by Pound. The results of these experiments are in full agreement with the theory of relativity. The present problem constitutes a simplified and schematized version of the idea of Pound's experiment.

9.34. Cerenkov radiation appears when the speed of light in the given medium is lower than the electron velocity. From the figure accompanying the answer we can see how the light wave is formed. In the time that it takes the electron to cover a path AB the light covers a distance AC, with

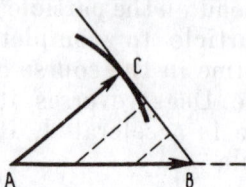

Fig. 9.34

$$\frac{|AC|}{|AB|} = \frac{c'}{v},$$

where $c' = c/n$ is the speed of light in the given medium. The envelope of the waves emitted by different points constitutes the wave front BC. The figure accompanying the answer shows that

$$\frac{|AC|}{|AB|} = \cos\theta.$$

The refractive index is

$$n = \frac{c}{v\cos\theta}.$$

Postface

Solution of the concluding problems in this *Collection* falls on the period when you are completing the general physics course in your college. It would be a mistake, however, to think that your studies in physics have come to an end. Physics will "pursue" you all your life unless, of course, you change your profession as engineer to that of opera singer or sports commentator.

Today numerous fields of human activity require a knowledge of physics, from astronautics to microbiology and from radio engineering to archeology.

But what portion of the physics studied in college will you find most needed in your future work? The laws? Naturally, one must know the main laws of physics, but I would not call this the most important aspect of your knowledge. The expression of a law or its mathematical formulation can be found in a reference book. This is even truer of the many specific formulas, such as the Poiseuille formula for viscous flow or the formula for the capacitance of a cylindrical capacitor.

Of course, the more formulas and laws that you remember the less frequently will you have to look into reference books and the more productive your work. And yet among the qualities that an engineer must have I would put first the ability to grasp the method required for a project. The aim of this book is to inculcate in the reader a taste for the physical method of thinking.

Solution of the majority of physical problems can be divided into four stages.

The first deals with the physical model of the phenomenon in question. A qualitative picture of the phenomenon is formulated, allowing for the factors that could be important. The second involves a mathematical model. An equation is set up that in accordance with an assumed law connects the factors introduced in the first stage. In the third stage mathematics steps in, so to say. By solving algebraic, trigonometric, or differential equations one can obtain the sought quantity in the form of an explicit function. The difficulties that arise in the third stage are more easily surmounted if the student has mastered the respective sections of mathematics. Mathematics for the engineer is what a cutting tool is for the lathe operator or a soldering iron for the assembler of electronic circuits.

Once the problem is solved, the very important fourth stage comes into the picture, namely, interpretation of the result obtained. The fourth stage is an analysis of the effect of the various parameters on the quantity of interest to the investigator.

To illustrate what has been said, let us examine damped oscillations, a common phenomenon known to everyone but not simple, nonetheless.

For instance, after performing several free oscillations, a pendulum finally stops; so does a load on a spring. The forces acting on the load are the elastic force exerted by the spring and the drag exerted by the surrounding medium (air). We assume that the elongation of the spring is small and, hence, the elastic force obeys Hooke's law. We also assume that the drag is proportional to the rate of motion of the load. All this constitutes the physical model of the phenomenon. Its mathematical model can be built by writing Newton's second law of motion: the mass of the load multiplied by the acceleration equals the sum of the projections of the forces acting on the load. This is a second-order differential equation, which can be solved (or integrated) if we consider the existence of two constants that depend on the initial data (the third stage). The resulting rather cumbersome formula expresses the time dependence of the load's displacement. The parameters in the formula are the mass of the load, the elasticity of the spring, and the resistance coefficient of the medium.

The analysis of the solution (the fourth stage) shows that a certain ratio of the parameters may produce periodic damped oscillations while another ratio may lead to aperiodic motion.

Such an analysis is given in a number of problems in this *Collection*. Take a careful look at their solution and pinpoint the four stages mentioned earlier. Give special consideration to the drawings accompanying the problems. Unfortunately, many students perceive a diagram as a simple illustration to be memorized and later drawn when necessary. As a result one sometimes gets a drawing that resembles a cartoon more than a physical diagram.

Often a student constructs the necessary curve more or less correctly but does not know the quantities that must be laid off on the axes. It is also difficult to overestimate the importance of knowing how to interpret a diagram. This requires, among other things, the skill of knowing how to "read" a diagram in the mathematical sense of the word, that is, understand that the derivative is positive where the curve goes up and negative where it goes down, and is zero at points of maxima and minima. In segments where the curve is convex downward the second derivative is positive; where it is convex upward the second derivative is negative. At inflection points the second derivative vanishes.

One must not forget that physics is an experimental science. In some cases an experiment helps one to find a sought law, discover a new phenomenon, or clarify certain aspects of a known effect; in others it serves as strict judge of the validity of a theory. Therefore, one must always prepare an experiment with care, understand the workings of the various devices involved, and analyze the results.

I believe that if you have solved or studied the solution of a large number of problems, the basics of the physical method of thinking have become clearer.

In conclusion I would like to hope that after you have finished college, far from being forgotten, physics will prove to be the real basis of your further development as an all-round person in this age of scientific and technical progress.